MYCOTOXIC FUNGI, MYCOTOXINS, MYCOTOXICOSES

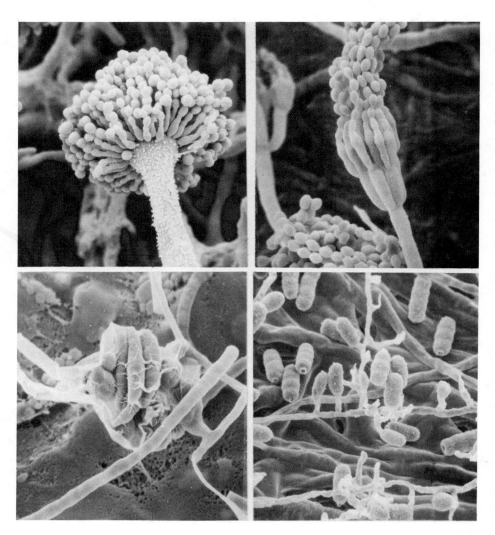

Scanning electron micrographs of four important mycotoxigenic fungi. From upper left clockwise: *Aspergillus flavus,* conidial head, 1050X; *Penicillium islandicum,* conidial head, 1400X; *Pithomyces chartarum,* conidiospores, 700X; *Fusarium tricinctum,* young macroconidia, 1750X. Photomicrographs courtesy of Dr. M. F. Brown, Department of Plant Pathology, University of Missouri-Columbia.

MYCOTOXIC FUNGI, MYCOTOXINS, MYCOTOXICOSES

An Encyclopedic Handbook

Volume 3

Mycotoxicoses of Man and Plants:
Mycotoxin Control and Regulatory Aspects

edited by

Thomas D. Wyllie
Department of Plant Pathology
College of Agriculture
University of Missouri–Columbia
Columbia, Missouri

Lawrence G. Morehouse
Department of Veterinary Pathology
and
Director of the Veterinary Medical Diagnostic Laboratory
College of Veterinary Medicine
University of Missouri–Columbia
Columbia, Missouri

MARCEL DEKKER, INC.   New York and Basel

Library of Congress Cataloging in Publication Data

Main entry under title:
Mycotoxicoses of man and plants.
    (Mycotoxic fungi, mycotoxins, mycotoxicoses; v. 3)
    Includes indexes.
    1. Mycotoxicoses. 2. Mycotoxins. 3. Food
contamination. 4. Plants, Effect of mycotoxins on.
5. Food adulteration and inspection–United States.
I. Wyllie, Thomas D. II. Morehouse, Lawrence G.
III. Series.
QR245. M88 vol. 3 [RA1250]   615. 9'52'92s   78–5168
ISBN 0–8247–6552–4   [614. 3'1]

MARCEL DEKKER, INC.

270 Madison Avenue, New York, New York 10016

Current printing (last digit):
10 9 8 7 6 5 4 3 2 1

PRINTED IN THE UNITED STATES OF AMERICA

FOREWORD

Mycotoxic fungi and their resulting threat to food and feedstuffs worldwide continue to have an extensive impact on the welfare of human and animal populations. Since early times these fungi have made their mark, but only lately have we been able to appreciate the extent of their impact. As recently as 1960, a major and widespread mycotoxicosis was initially thought to be a strange, new "X disease," probably of viral origin. The discovery of the role of peanuts in an aflatoxicosis was the leading factor contributing to increased interest and concern in the United States and, indeed, worldwide. In the ensuing 17 years there has been a more organized effort devoted to understanding and, hopefully, controlling the problem created by these fungi.

The range of food and feedstuffs potentially affected by mycotoxic fungi and the numbers of microbial species has been extensively increased in recent years. Although the epidemic occurrences seem to have lessened, the endemic and more subtle effects, no less destructive, appear to be increasing. In part this results from an awareness brought about by new knowledge and the ability to document cause and effect relationships.

A wealth of knowledge about mycotoxic fungi, their toxins, and the resulting toxicoses has been documented in scientific literature throughout the world. This encyclopedic handbook brings this large volume of knowledge together through the collective efforts of experts analyzing the documentation in every facet of the field. From the isolation, culture, and identification of suspect or known myco-toxic fungi, to the chemistry of toxins and toxicoses, the subjects are synthesized and set forth for the reader in an orderly, accessible manner. The organization and indexing are designed to make these three volumes readily usable as standard, working references. Their comprehensiveness and international scope of subject coverage makes them an invaluable reference for students, scientists, physicians, veterinarians, animal health specialists, public health and regulatory personnel, and many others involved with the planning, production, and use of food and feed-stuffs.

John F. Fulkerson
Principal Scientist
Cooperative State Research Service
United States Department
of Agriculture
Washington, D. C.

This three-volume Encyclopedic Handbook of mycotoxic fungi, mycotoxins, and mycotoxicoses assembles information on the fungi, their toxic metabolites, and the diseases associated with them. Well-documented mycotoxicoses are discussed, and comment is made on less well-known disease syndromes suspected to have a mycotoxic origin. There are 11 countries represented by 46 contributing authors. This affirms the international scope and nature of the mycotoxin problem, and emphasizes the importance of specific disease syndromes in certain locales, while suggesting that mycotoxic fungi known to cause disease in one country may be present in another with only negligible impact on animals and man. From such information, we infer that the factors responsible for the production and consumption of toxic compounds probably are controllable through a shift in agricultural procedures or food processing, marketing, shipping, and storage practices. These approaches are well-documented and generally recognized as effective ways to eliminate mycotoxin problems. Less often considered is the contribution to control of the problem by changing current production methods.

This Handbook has been designed to afford an effective introduction to, and review of, the complex field of mycotoxicology. We were especially concerned that the research scientist, practicing veterinarian, and physician as well as individuals responsible for the production and safeguarding of our food supply be able to locate readily pertinent information about the subject. The book is therefore organized into sections on mycotoxic fungi, specific toxins, affected animal species, and is extensively cross-referenced.

The Handbook collates information pertaining to the identification and manipulation of mycotoxic fungi, chemistry of toxic compounds, and their physical and chemical properties. Methods of their extraction, isolation, and identification

v

are defined.   Metabolism of the toxins and their effect on man, domestic and laboratory animals, poultry, aquatic species, and plants is described.   In addition, gross
and cellular pathology, as well as possible control or treatment of the diseases
they cause, are discussed in detail.

Thomas D. Wyllie
Lawrence G. Morehouse

INTRODUCTION

An association of toxic substances with plants and plant products used for human
or animal consumption has been made since biblical times, with ergotism representing one of the first suspected effects of such compounds. In modern times
discussions of mold-produced poisons appeared in 1924 concerning Aspergillus
toxicity in cattle; in 1931 on stachybotryotoxicosis in horses; in 1936 on Fusarium
toxicity and digestive disturbances in swine; in 1940 on moldy corn poisoning in
horses; in 1943-44 on alimentary toxic aleukia caused by the consumption of
Fusarium contaminated cereals in the U.S.S.R. In 1954, production of clinical
and pathologic symptoms of "X disease" in cattle associated with toxic Aspergillus
spp. was reported, and in the mid and late 1950s there were reports of disease of
cattle and swine caused by feeding moldy corn. Since 1960, when "Turkey X disease" was reported, incriminating aflatoxin in its etiology, numerous studies have
been conducted on the nature of the "aflatoxin family" of compounds and the various
clinical problems encountered as a result of their consumption. The impact of
the discovery of "aflatoxicosis" as a distinct disease syndrome, and the subsequent
implication of aflatoxins as potential carcinogens in the human food chain, was
felt on an international scale. It rapidly became a matter of great concern to
scientists and others interested in the production, manufacturing, and handling of
food and feed products, and to livestock and poultry producers. It is considered a
potentially major threat to public health. Although research with aflatoxin has received major emphasis and this effort continues, attention has now also turned to
other mycotoxins concerning their cause and effect relationship with known disease

syndromes or their possible implication in a number of idiopathic diseases. Hence, the concern for "aflatoxin" and "aflatoxicoses" has become a broader one for "mycotoxins" and "mycotoxicoses."

A wide variety of fungi and fungal metabolites have come under scrutiny. Findings of recent years have far-reaching implications that include not only the quality of food of both plant and animal origin, but also the availability of "clean products." Aspects of this subject were discussed in the recent symposium on mycotoxins during the 1975 Annual Meeting of the American Society for Microbiology and are published in Microbiology - 1975 (edited by David Schlessinger). The influence of "naturally occurring" mycotoxin residues has had a major impact on food industries of the world and is reflected in the controlled movement of some food and feed foodstuffs across international borders. The ultimate concern is that some of the mycotoxins are carcinogenic, and thus human health may be directly threatened. Although emerging and underdeveloped nations of the world may be more concerned with the simple availability of food of any quality, in the United States the concern for a loss of quality in the food supply is reflected in the food additives amendment of the Federal Food, Drug, and Cosmetic Act, Chapter IV, section 402 (a)l. Further deliberations of the zero tolerance policies on possible carcinogens in the food chain likely will depend, at least in the United States, on the sophistication of analytic methods of detection balanced against the wisdom of our scientists and lawmakers to determine reasonable legal limits for mycotoxins in foodstuffs and feedstuffs. Other concerns include the influence of mycotoxins on recreation and conservation interests as some of these substances involve fish and wildlife. Little information is available concerning the potentiation of mycotoxins by other mycotoxins, antibiotics, pesticides, and other environmental pollutants.

It was apparent to us that despite the many scientific reports, monographs, symposia, and textbooks on various aspects of mycotoxins and mycotoxicoses, a need existed to bring together in a single work, current information on a wide range of toxic fungal metabolites, the fungi that produce them, and the resultant diseases they cause. Therefore, our major objective was to organize the Handbook for maximum ease of cross-referencing between the parts devoted to mycology, chemistry of toxins, and the disease syndromes elicited by them. Each contributor was given license to explore a subject as he willed and in his own method of expression, and to assume responsibility for reviewing and citing pertinent literature in his area of expertise. Thus, the uniformity of format would be attained but individual expression by authors would remain. Furthermore, the objective of a complete reference work as well as a practical bench-level guide could be achieved.

ORGANIZATION OF THE HANDBOOK

The Handbook is divided into three volumes:

     Volume 1   Mycotoxic Fungi and Chemistry of Mycotoxins
     Volume 2   Mycotoxicoses of Domestic and Laboratory Animals,
                    Poultry, and Aquatic Invertebrates and Vertebrates
     Volume 3   Mycotoxicoses of Man and Plants: Mycotoxin Control
                    and Regulatory Practices

(see Contents of this volume and Contents Of Other Volumes for further details). The Handbook is composed of six parts and supporting subsections that provide ancillary information on concepts and information of a more generalized nature. The first is concerned with mycology. It is designed to acquaint the uninitiated with the characteristics of the mycotoxic fungi. It permits one to determine whether an isolate obtained from suspect feed or foodstuffs could be a mycotoxic form.

Part 1 begins with a comprehensive key to the genera of fungi that have been reported as producers of mycotoxins regardless of their significance in the causation of disorders of animals or man. Therefore, several genera are included that are not widely recognized as "mycotoxic" fungi. The genera covered are illustrated for ease of identification. Subsequent subsections deal with keys to the species of the major "mycotoxic" genera. In general, the authors have made statements about the types of substrate and environmental conditions that favor growth of an organism, production of its characteristic toxins, and the animals affected by them.

Taxonomic treatment of any group of fungi rarely is accepted universally. Usually, the controversies surrounding a particular group are not overly severe. Treatment of the genus _Fusarium_, however, is another matter. Several attempts have been made to classify the fusaria over the years. Some treatments are cumbersome and difficult to use; other are usable, but oversimplified in the view of many experts. In fact, no single taxonomic treatment is universally accepted today, and we suspect that there will be additional attempts to classify this group of fungi in the future. The treatment used in this Handbook is an attempt to select a middle-of-the-road approach to the fusaria. The treatment is more detailed than the Snyder and Hansen classification which is generally accepted by the plant pathology community. However, in the view of the author of this section, the Snyder and Hansen classification was based primarily on the plant pathogenic properties of the fusaria and not on their mycotoxigenic properties, and therefore we considered it to be inappropriate for the Handbook. The present treatment allows for a multidisciplined recognition of the fusaria to species as they appear in the international literature, while at the same time allowing those preferring a more simplified taxonomy to utilize the key. All synonomies between the various classifications are included in the detailed description of the species to facilitate recognition by the researcher regardless of taxonomic preference.

With respect to the treatment of the major mycotoxic genera, each genus is keyed to species but no attempt is made to include all species in a given genus. Only those recognized, reported, or suspected of being mycotoxigenic are included.

Part 2 is concerned with the chemistry of mycotoxins. It is designed as both a bench-level guide and as a reference source. Our intention was to assemble specific information on the physical and chemical properties of the toxins, methods of extraction, isolation, purification, and identification and to give full treatment to the biochemistry and physiologic effects of each toxin. Entrance to more indepth consideration of a topic is provided by very detailed referencing on each subject covered. All references are fully titled for the benefit of the reader in evaluating the pertinence of each individual reference to his interest. This section is organized by toxin, and aside from the chemical information presented, information concerning the fungi involved and the animal species affected is given. The section is fully cross-referenced into Part 1 (Mycology of Mycotoxic Fungi) and Part 3 (Mycotoxicoses) for rapid access to supporting information in those fields.

Part 3 is concerned with the mycotoxicoses. It was specifically designed to offer information about mycotoxicoses to the research scientist, practicing veterinarian and physician, and individuals interested in livestock, poultry, and aquatic animal production. The section is organized by animal species. Part 3, Section 1 presents an overview of diagnostic procedures for mycotoxicoses in animals, followed by Sections 2 through 6 on mycotoxicoses of cattle, sheep, horses, swine, and poultry, respectively. Hence individuals who specialize in given categories of animal species, either in production, disease therapy, or in specific research areas may gain a rapid overview as well as an in-depth insight into the mycotoxicoses that have been reported in that particular species. This section is also cross-referenced to the causal fungi and specific toxins involved.

The subject matter covered by animal species includes information on the pathology, clinical symptomatology, occurrence, and potentiating factors and control or treatment for each disease. The study of mycotoxicoses is an emerging science, and there is obviously far more known about some than others. However, in this Handbook we speak to the reported mycotoxicoses and indicate the extent of the knowledge at this time. Some redundancy was unavoidable, but where it occurs, it has been permitted with the intent of allowing each section to stand independently without referring the reader to a wider spectrum of disease or species of animal than necessary.

Part 3, Section 7 represents the major coverage on laboratory animals, although the Handbook contains other discussions by some authors who have specific experience with certain mycotoxins on individual laboratory animal species. The contribution in Section 8 covers the range of available knowledge on mycotoxicoses in the aquatic species, including both vertebrates and invertebrates. It also is organized on a specific species basis with each toxin covered.

Part 3, Sections 9 through 13 deal with mycotoxicoses of man and specific conditions in man known or suspected of being associated with the consumption of mycotoxins. The section is not organized strictly on a "system basis." However, reference is made to the central nervous and digestive systems, i.e., aflatoxicosis and alimentary toxic aleukia; skin, i.e., toxic verrucarins and roridins, sporides-

min, ergotism, stachybotryotoxicosis; cardiovascular system, i.e., cardiac beriberi, and general effects of mycotoxins on vascular permeability; hemic and
lymphatic system, i.e., bone marrow depressant effects; urinary system, i.e.,
possible role of ochratoxins in nephropathies of man. Section 12 speaks to the
role of mycotoxins in human pulmonary disease. An introduction to the overall
section on human mycotoxicoses covers the dietary and epidemiologic considera-
tions of the mycotoxic problem in humans.

We believe that the diseases described and referenced in the Handbook represent a comprehensive treatment of mycotoxic fungi and mycotoxicoses. We realize
that a broader range of idiopathic disease syndromes may be ascribed to the mycotoxins in the future as our knowledge of the fungi, the toxins, and the diseases increase. It is also emphasized that the intent in this work was to deal with the effect of a single mycotoxin on a single species. However, a variety of fungi, producing a variety of compounds, acting in concert, may be involved in the diseases
described herein. Inferences to this effect are made in various chapters of the
Handbook where it seemed appropriate. It is further realized that enhancement of
mycotoxicoses by other disease organisms, or vice versa, may occur, and evidence
of this is beginning to appear in the literature. Studies suggest that some agents
may predispose organisms to general debilitating effects and increase susceptibility or sensitivity to disease incited by bacteria, viruses, or toxins.

Part 4 is concerned with the effects of the mycotoxins on higher plants, bacteria,
and algae, thus completing the review of the effects of mycotoxins on animal and
plant life.

Part 5 is concerned with control procedures for mycotoxic fungi. The growth
of these fungi both in the field and in storage is discussed. The manner in which
fungal growth and subsequent toxin production can be restricted or inhibited is out-
lined. Facts related to the utilization or salvage of the product are given. The
section briefly summarizes the general concepts alluded to by many of the authors
during their discussion of various specific fungi, substrates, and uses, regarding
methodology related to their manipulation.

Part 6, the concluding chapter of the Handbook, is devoted to regulatory aspects
of the mycotoxin problem in the United States. Mycotoxins currently controlled by
law are discussed as well as some indication of where regulation of mycotoxins in
food and feedstuffs may be headed in the future.

In summary, we trust that this format has provided an access to the field of
mycotoxins for mycologists, plant pathologists, biologists, food scientists, chemists
and biochemists, toxicologists, veterinarians, physicians, and others interested in
livestock and poultry production and aquatic species. Individuals in various scientific disciplines can refer immediately to their area of interest for information
and for rapid guidance to other areas of concern. Supporting sections include procedures for diagnosing a mycotoxicosis, the current concepts of proper storage
facilities, of contamination of raw food and feedstuffs by mycotoxins, control and
marketing implications, and current legal and FDA regulations concerning mycotoxins in their food supply.

The editors acknowledge the support of USPH Biomedical Research Support
Grants Nos. 1624 and 1625 awarded by the University Research Council. We also
acknowledge the helpful suggestions of many colleagues at the University of
Missouri - Columbia while this work was being organized and assembled. We
are particularly grateful to Professor Anton Novacky, Department of Plant Pathology, for his translation of the Russian articles; to Professors D.A. Schmidt and
A.W. Hahn, College of Veterinary Medicine, for their assistance in interpretation
of clinical pathologic and electrocardiographic data, respectively, translated from
Russian; to Mary Allcorn and Georgia Morehouse for their assistance with subject
indexing; to Donald Connor, Artist, College of Veterinary Medicine, for the
cover design; and to the many staff members at Marcel Dekker, Inc. with whom we
have enjoyed working and for their excellent copyediting and attention to detail
which has greatly simplified our task. Most importantly, we wish to thank the
contributors for their cooperation in making this handbook possible. The ultimate
usefulness of the work resides with each of them.

Thomas D. Wyllie
Lawrence G. Morehouse

CONTENTS OF VOLUME 3

Foreword   iii
Preface   v
Introduction   vii
Contributors to Volume 3   xix
Contents of Other Volumes   xxi

PART 3 (continued)
MYCOTOXICOSES

3.9   Mycotoxicoses of Man: Dietary and Epidemiological Conditions   1
          (Ronald C. Shank)

          3.9.1   Introduction   1
          3.9.2   Historical Background   1
          3.9.3   Human Aflatoxicoses   6
          3.9.4   Problems in Epidemiological Design   13
          3.9.5   Summary   15

3.10   Fusarium Poae and F. Sporotrichioides as Principal Causal
          Agents of Alimentary Toxic Aleukia   21
          (Abraham Z. Joffe)

          3.10.1   Introduction   21
          3.10.2   Epidemiological Background   21
          3.10.3   Bioassay Methods and Procedures   26
          3.10.4   Fungi and Toxins   30
          3.10.5   Etiology   41
          3.10.6   Clinical Characteristics of Alimentary Toxic
                     Aleukia   54
          3.10.7   Pathology of Organs in Man and Animals   59
          3.10.8   Prophylaxis and Treatment   70
          3.10.9   Summary   70
          3.10.10   Recent Developments   71

3.11   Human Stachybotryotoxicosis   87
          (E.-L. Hintikka — née Korpinen)

          3.11.1   Introduction   87
          3.11.2   Organ Systems Affected and Symptomatology   87
          3.11.3   Control   88

3.12   The Role of Mycotoxins in Human Pulmonary Disease   91
          (Sharon C. Northup and Kaye H. Kilburn)

          3.12.1   Introduction   91
          3.12.2   Types of Mycotoxic Pulmonary Disease   92
          3.12.3   Pulmonary Zones Affected by Fungi   93
          3.12.4   Mechanisms of Diseases Due to Fungi   95
          3.12.5   Experimental Evidence for Pulmonary Toxicity by
                      Mycotoxins   96
          3.12.6   Conclusions   104

3.13   Possible Role of Mycotoxins in the Hemic System in Man   109
          (El-Sheikh Mahgoub)

PART 4
EFFECTS OF MYCOTOXINS ON HIGHER PLANTS, ALGAE, FUNGI,
AND BACTERIA
(Jürgen Reiss)

4.1   Introduction   119
4.2   Aflatoxins   119
4.3   Patulin   127
4.4   Rubratoxin B   129
4.5   Byssochlamic Acid   131
4.6   Diacetoxyscirpenol   131
4.7   Other Fusarium Mycotoxins   132
4.8   Cytochalasins   133
4.9   Other Mycotoxins   134

PART 5
CONTAMINATION BY MYCOTOXINS: WHEN IT OCCURS AND HOW TO
PREVENT IT
(David B. Sauer)

5.1   Introduction   147
5.2   Mycotoxins Produced in the Field   147
5.3   Aspergillus flavus Development in the Field   148
5.4   Contamination by Mycotoxins During Storage   150
5.5   Grain Drying   152

5.6   High Moisture Grain Storage   154
5.7   Summary   155

PART 6
REGULATORY ASPECTS OF THE MYCOTOXIN PROBLEM IN THE
UNITED STATES
(Joseph V. Rodricks)

6.1   Introduction   161
6.2   General Characteristics of the Problem of Mycotoxins in
          Foods   161
6.3   Evolution of the Current Aflatoxin Control Program   163
6.4   Research Activities Related to Mycotoxin Control   167
6.5   Development of Programs for Other Mycotoxins   168
6.6   Future Trends   169

Glossary   173
Author Index   183
Subject Index   195

CONTRIBUTORS TO VOLUME 3

EEVA-LIISA HINTIKKA (née Korpinen)   Department of Microbiology and Epizootology, College of Veterinary Medicine, Helsinki, Finland

ABRAHAM Z. JOFFE   Department of Botany, Laboratory of Mycology and Mycotoxicology, The Hebrew University of Jerusalem, Jerusalem, Israel

KAYE H. KILBURN*   Department of Medicine, School of Medicine, University of Missouri-Columbia, Columbia, Missouri

EL-SHEIKH MAHGOUB[†]   Department of Microbiology and Parasitology, Faculty of Medicine, University of Khartoum, Khartoum, Sudan

SHARON NORTHUP[††]   Department of Research Service, Veterans Administration, Truman Memorial Hospital, Columbia, Missouri

JÜRGEN REISS   Mikrobiologisches Laboratorium, Bad Kreuznach, Germany

JOSEPH V. RODRICKS   Bureau of Foods, Food and Drug Administration, Department of Health, Education and Welfare, Washington, D.C.

DAVID B. SAUER   U.S. Grain Marketing Research Center, Agricultural Research Service, U.S. Department of Agriculture, Manhattan, Kansas

RONALD C. SHANK[¶]   Department of Nutrition and Food Science, Massachusetts Institute of Technology, Cambridge, Massachusetts

*Current affiliation: Departments of Medicine and Community Medicine, Mt. Sinai School of Medicine, The City College of the City University of New York, New York, New York

[†]Current Affiliation: Department of Microbiology and Parasitology, Faculty of Medicine, Riyadh University, Riyadh, Saudi Arabia

[††]Current affiliation: Travenol Laboratories, Morton Grove, Illinois

[¶]Departments of Community and Environmental Medicine, and Medical Pharmacology and Therapeutics, College of Medicine, University of California at Irvine, Irvine, California

CONTENTS OF OTHER VOLUMES

VOLUME 1
MYCOTOXIC FUNGI AND CHEMISTRY OF MYCOTOXINS

Part 1   Mycology of Mycotoxic Fungi

1.1   A Key to the Genera and Selected Species of Mycotoxin-Producing
          Fungi
          (E. E. Butler and E. V. Crisan)

1.2   The Genus Aspergillus
          (John Tuite)

1.3   The Genus Penicillium
          (Philip B. Mislivec)

1.4   The Genus Fusarium
          (Abraham Z. Joffe)

1.5   The Genus Claviceps
          (P. G. Mantle)

1.6   The Genus Stachybotrys
          (E.-L. Hintikka — née Korpinen)

1.7   The Genus Pithomyces
          (M. E. di Menna, P. H. Mortimer, and E. P. White)

1.8   The Genus Phoma
          (M. E. di Menna, P. H. Mortimer, and E. P. White)

1.9   The Genus Myrothecium
          (M. E. di Menna, P. H. Mortimer, and E. P. White)

1.10   The Genus Phomopsis
          (W. F. O. Marasas)

1.11   The Genus Diplodia
          (W. F. O. Marasas)

Part 2   Chemistry of Mycotoxins

2.1   Aflatoxin and Related Compounds
           (D. S. P. Patterson and B. D. Jones)

2.2   Chemistry of Coumarin and Related Compounds
           (Tsune Kosuge and D. Gilchrist)

2.3   Penicillium Mycotoxins
           (Peter M. Scott)

2.4   Chemistry of the Tremorgenic Mycotoxins
           (S. J. Cysewski)

2.5   Chemistry of Fusarium and Stachybotrys Mycotoxins
           (Chester J. Mirocha, Sadanand V. Pathre, and Clyde M. Christensen)

2.6   Chemistry of Claviceps Mycotoxins
           (P. G. Mantle)

2.7   Chemistry of the Sporidesmins
           (E. P. White, P. H. Mortimer, and M. E. di Menna)

2.8   Chemistry and Physiology of Slaframine
           (E. B. Smalley)

2.9   Toxins of Phoma herbarum var. medicaginis
           (E. P. White, P. H. Mortimer, and M. E. di Menna)

2.10   Chemistry of the Myrothecium Toxins
           (E. P. White, P. H. Mortimer, and M. E. di Menna)

VOLUME 2
MYCOTOXICOSES OF DOMESTIC AND LABORATORY ANIMALS, POULTRY,
AND AQUATIC INVERTEBRATES AND VERTEBRATES

Part 3   Mycotoxicoses

3.1   Diagnosis of Mycotoxicoses in Animals and Poultry: An Overview
           (G. H. Nelson and Clyde M. Christensen)

3.2   Mycotoxicoses in Cattle
           (C. N. Cornell, S. J. Cysewski, M. E. di Menna, George B.
           Garner, E.-L. Hintikka (née Korpinen), T. S. Kellerman, Alexander
           C. Keyl, I. A. Kurmanov, P. G. Mantle, W. F. O. Marasas, P. H.
           Mortimer, William E. Ribelin, L. D. Scheel, E. B. Smalley,
           E. P. White)

3.3   Mycotoxicoses in Horses
           (Charles H. Bridges, E.-L. Hintikka (née Korpinen), P. G. Mantle,
           W. F. O. Marasas)

3.4   Mycotoxicoses in Sheep
        (Bernard H. Armbrecht, M. E. di Menna, E.-L. Hintikka (née
        Korpinen), I. A. Kurmanov, P. G. Mantle, W. F. O. Marasas,
        P. G. Mortimer, E. P. White)

3.5   Mycotoxicoses in Swine
        (Bernard H. Armbrecht, E.-L. Hintikka (née Korpinen), Palle
        Krogh, Harold J. Kurtz, P. G. Mantle, W. F. O. Marasas,
        Chester J. Mirocha)

3.6   Mycotoxicoses in Poultry
        Peter K. C. Austwick, E.-L. Hintikka (née Korpinen), Abraham Z.
        Joffe, I. A. Kurmanov, P. G. Mantle, John C. Peckham)

3.7   Mycotoxicoses in Laboratory Animals
        (William W. Carlton, S. J. Cysewski, M. E. di Menna, E.-L.
        Hintikka (née Korpinen), W. F. O. Marasas, P. H. Mortimer,
        G. M. Szczech, E. P. White)

3.8   The Effect of Mycotoxicoses in Aquatic Animals
        (R. O. Sinnhuber and Joseph H. Wales)

MYCOTOXIC FUNGI, MYCOTOXINS, MYCOTOXICOSES

PART 3 (continued)

MYCOTOXICOSES

3.9 Mycotoxicoses of Man: Dietary and Epidemiological Considerations

3.9.1 Introduction

Most of the mycotoxin literature concerning toxicity has focused on veterinary problems for two reasons: (a) in terms of frequency, animal intoxications are of overriding importance; most veterinary outbreaks have been acute poisonings, in many instances not difficult to recognize, whereas indications of man suffering from acute mycotoxicoses have been extremely rare; (b) it is relatively simple to test a suspected mycotoxin etiology in an outbreak of livestock poisoning by administering known amounts of the toxin to healthy animals under controlled conditions.

There are only three mycotoxicoses for which there exists reasonable evidence to associate the toxins with human disease. These are the well-known disease of ergotism, alimentary toxic aleukia (ATA) of Russia, and liver cancer (and possibly Reye's syndrome), presumably related to aflatoxin consumption. This section reviews the human disease aspects of ergotism and ATA and then analyzes epidemiological evidence from four studies that attempted to correlate aflatoxin exposure to human liver cancer. The relationship between aflatoxin and both Indian childhood cirrhosis and Reye's syndrome is also discussed. Efforts to determine the role of any mycotoxin in human disease will usually employ epidemiological techniques, and therefore the section closes with comments on some of the major problems in the design of these association studies in human disease.

3.9.2 Historical Background

3.9.2.1 Ergotism

The oldest and best known human mycotoxicosis is ergotism. Centuries ago in Central Europe a disease occurred irregularly and was characterized by a sensation of cold hands and feet which progressed to an intense burning sensation. In advanced cases the extremities became gangrenous and necrotic. Outbreaks of ergotism reached epidemic proportions in the Middle Ages, and it was thought in those times that a trip to the distant shrine of St. Anthony would bring about relief from the intense burning sensations; hence, the disease became known popularly as St. Anthony's fire. The curative effect of the trip to the shrine perhaps resulted from removal of the victim from the area of

1

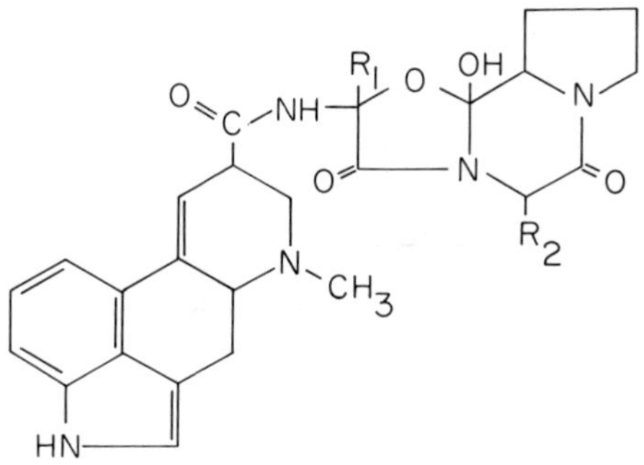

Fig. 1   General structure for ergot alkaloids derived from lysergic acid.
Radicals are defined in Table 1 for ergotamine, ergosine, ergostine, ergo-
cristine, α-ergokryptine, β-ergokryptine, and ergocornine.

environmental contamination by the causative agent.   Bové [13] has written an
excellent treatise on the pharmacognosy, chemistry, and physiology of
ergotism.

   The disease became associated with the consumption of bread made from
flours of rye and other grains overgrown with toxigenic strains of the molds
Claviceps purpurea and C. paspali (Sec. 1.5).   The molds developed only in the
female sex organs of the grasses, producing ergots, black or dark purple
compact masses of hardened mycelium [54].   It was not until the seventeenth
century that it was recognized that alkaloids produced by the ergot were
responsible for the disease, even though these alkaloids had long been known
as powerful oxytocics.

   These compounds are derivatives of ergoline and can be classified as
lysergic acid derivatives (see also Sec. 2.6.3) or clavine alkaloids.   The
lysergic acid compounds are represented in Figs. 1 and 2, and the radicals

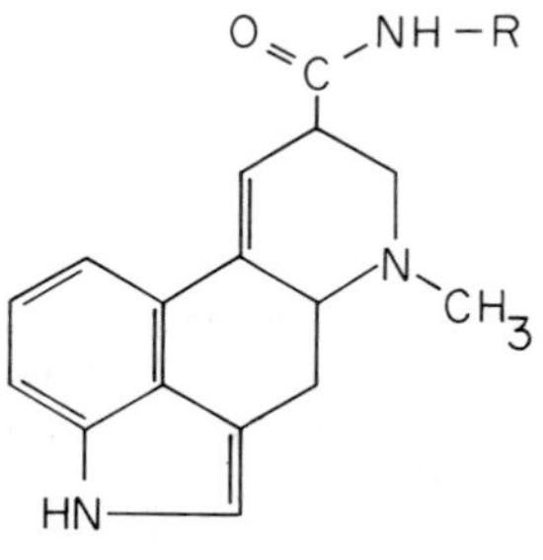

Fig. 2   General structure for peptide-type ergot alkaloids derived from lysergic acid.   Radical is defined in Table 2 for ergine, lysergic acid methylcarb-
inolamide, ergometrine, and lysergic acid-L-valinemethylester.

Table 1   Ergot Alkaloids Derived from Lysergic Acid[a]

Alkaloid	R_1	R_2
Ergotamine	CH_3-	$C_6H_5CH_2-$
Ergosine	CH_3-	$(CH_3)_2CH-CH_2-$
Ergostine	CH_3CH_2-	$C_6H_5CH_2-$
Ergocristine	$(CH_3)_2CH-$	$C_6H_5CH_2-$
α-Ergokryptine	$(CH_3)_2CH-$	$(CH_3)_2CH-CH_2-$
β-Ergokryptine	$(CH_3)_2CH-$	$C_2H_5CH(CH_3)-$
Ergocornine	$(CH_3)_2CH-$	$(CH_3)_2CH-$

[a]Radicals refer to Fig. 1.

R_1, R_2, and R are defined in Tables 1 and 2. The biologically active ergot alkaloids are derivatives of d-lysergic acid and are designated by the suffix -in. The dextrorotatory derivatives of d-isolysergic acid are designated by the suffix -inine, and are only weakly bioactive.

Most textbooks of pharmacology present extensive discussions of the biological activity of the ergot alkaloids. These compounds produce α-adrenergic blockade, inhibiting certain responses to adrenergic nerve activity, to epinephrine, and to 5-hydroxytryptamine. They bring about marked peripheral vasoconstriction, which if not corrected, can result in gangrene. The compounds are also highly active in direct stimulation of smooth muscle and have been used as oxytocics to increase the force and/or frequency of uterine contractions. Ergot alkaloids also have effects on the central nervous system which include stimulation of the hypothalamus and other sympathetic portions of the midbrain and depression of the vasomotor center; these compounds are also centrally acting emetics. Instead of causing human suffering, several of the alkaloids are now widely used to treat human diseases.

Table 2   Peptide-Type Alkaloids Derived from Lysergic Acid[a]

Alkaloid	R
Ergine	H
Lysergic acid methylcarbinolamide	$CH_3CH(OH)-$
Ergometrine (ergobasine)	$CH_2OH-CH(CH_3)-$
Lysergic acid-L-valinemethylester	$(CH_3)_2CH-CH(COOCH_3-$

[a]Radical refers to Fig. 2.

3.9.2.2   Alimentary Toxic Aleukia
           (ATA)

Outbreaks of a disease affecting the hematopoietic system resulting in
decreased formation of red and white blood cells and platelets have occurred
in Russia [21] (see also Sec. 3.10 of this text).   Large numbers of people
were affected and mortalities were as high as 60%.   The syndrome consisted of
fever, hemorrhagic rash, bleeding from the nose, throat, and gums, necrotic
angina, extreme leukopenia, agranulocytosis, sepsis, and exhaustion of bone
marrow.   Joffe [21] cites Russian workers who suggested that ATA results
from indirect toxic injury to the hematopoietic, autonomic nervous, and endocrine systems.   Mayer [27, 28] presented a comprehensive review of several
aspects of the disease.

The clinical course of ATA follows four stages [27]: (a) Within hours of
eating contaminated bread a mild inflammation from the mouth to the stomach
occurs and develops into acute gastroenteritis; this local effect persists several
days before ending spontaneously, even with continued exposure; (b) a period of
several weeks follows in which there are few gross symptoms but the bone
marrow undergoes destruction; after extensive damage hemorrhagic spots
appear on the skin and there is a sharp decrease in the number of leukocytes;
disturbances of the central and autonomic nervous systems are apparent,
especially weakness, vertigo, headache, palpitation, slight asthma, and
lowered blood pressure; (c) severe atrophy of the bone marrow with hemorrhagic diathesis, necrotic angina, sepsis, and marked decrease in the numbers
of leukocytes, granulocytes, erythrocytes, and platelets; (d) either death or a
period of recovery.

Outbreaks of the disease have been associated with ingestion of overwintered grains which became infected with toxigenic species of Fusarium
(Sec. 1.4.6).   It was customary in Russia to harvest after the crops had been
left in the fields all winter, and under the conditions of repeated freezing and
thawing, molds produced several toxic metabolites.   Olifson [32] reported
isolating two toxic glucosides from prosomillet infested with Fusarium which
had been isolated from overwintered grain.   Fusarium sporotrichiodes (Sec.
1.4.6.3) and F. poae (Sec. 1.4.6.1) produced sporofusarin and poaefusarin,
respectively; when these compounds were given to cats they produced a syndrome similar to ATA in man.

These same molds are producers of the toxic 12,13-epoxytrichothecenes,
several of which have been shown to be radiomimetic [37, 47, 48].   Recently,
Mirocha and Pathre [29] reported that a sample of poaefusarin contained a
sufficient amount of the epoxytrichothecene, T-2 toxin (Sec. 2.5.1.3), to fully
account for the sample's toxicity, thus inferring that T-2 toxin, not the fusarin
glucosides, might be the toxic agent in ATA (Fig. 3).   Bamburg and co-workers
[7] in 1969 suggested that epoxytrichothecenes might be responsible, at least
in part, for this toxicosis, and Wyatt and co-workers [51, 52] have pointed out
similarities between T-2 poisoning in chickens and ATA in man.   The etiology
of this toxic grain disease is complex and a variety of factors may be critical.
Mayer [27] points out that the Russian victims were often quite poor and were
malnourished or undernourished for prolonged periods.

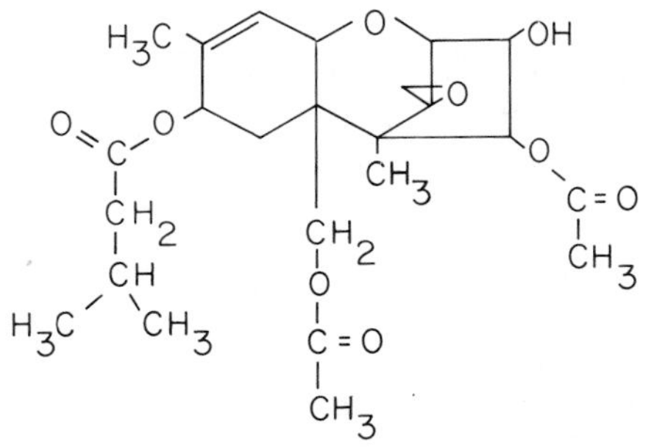

Fig. 3   Structure for 4β-15-diacetoxy-8α-(3-methylbutyryloxy)-12,13-epoxy-trichothec-9-en-3α-ol, T-2 toxin.

3.9.2.3   Acute Cardiac Beriberi

In the latter half of the nineteenth century epidemics of an acute heart disease broke out in rural Japan. After a detailed retrospective analysis of cases of the disease termed acute cardiac beriberi (shoshin-kakke), Uraguchi [49] concluded that the disease probably represented a human mycotoxicosis.

Acute cardiac beriberi is not an avitaminosis but was termed so because of its association with the consumption of polished rice. In 1910 the number of acute cardiac beriberi cases suddenly decreased, yet polished rice remained prominent in the diet and other forms of beriberi did not decline; also in that year the Japanese government implemented an accelerated program of rice inspection to reduce the amount of moldy grain getting to the markets, which perhaps could explain the end of the epidemics if mycotoxins were involved.

The disease is characterized by precordial distress with palpitation and tachypnea; as the dyspnea worsens nausea and vomiting are experienced. After a few days the victim suffers severe anguish, pain, and restlessness and at times can become violently maniacal. The right heart is dilated and the heart sounds are abnormal. In the last stages, as the dyspnea increases, the extremities become cold and cyanotic, the pupils dilate, and the person loses consciousness. Uraguchi [49] has suggested that the neurotoxic mycotoxins of Penicillium citreo-viride Biourge (Sec. 1.3.4.1) may be causally related to acute cardiac beriberi, but, lacking an animal model for the disease and having to rely wholly on retrospective studies of human cases, there is no means with which to test this hypothesis.

3.9.3   Human Aflatoxicoses

3.9.3.1   Chronic Toxicity of
           Aflatoxin B_1

The aflatoxins have been discussed in earlier sections. Especially important
in the development of the aflatoxin problem were two factors: widespread contamination by the toxins in man's foods and foodstuffs, and the intense
carcinogenicity of aflatoxin B_1 (Sec. 2.1.2) in experimental animals. Only 2
years after their discovery, LeBreton and co-workers [24] suggested that the
aflatoxins may be a causative factor in human liver cancer. Hepatocellular
carcinoma is unusually high in sub-Sahara Africa and in Southeast Asia, and
after studying the evidence for mycotoxin contamination of man's food supply
and the geographical distribution of liver cancer in the world, Oettlé [31] and
Kraybill and Shimkin [23] concluded that dietary aflatoxins could be related to
human liver cancer. A series of international field studies to test this
proposed relationship was then initiated.
    The first of these studies was carried out in Uganda from 1964 to 1967.
Hepatoma incidence was determined from cancer registry data for 1964 to 1966,
a total of 355 cases from a general population of 5.8 million. The accuracy of
the clinical diagnosis was estimated to be 85% [3]. The incidence was highest
in the age range of 35 to 45 years and was 3.3 times greater in males; in the
21 tribes studied there were approximately two new cases of hepatoma per
100,000 people per year except in the Rwanda, the Karamojong, and the Suk,
in which the incidence was 3.0, 3.5, and 10.7, respectively [2].
    Food samples were collected from home granaries or village markets
from September 1966 to June 1967 [4] and assayed for aflatoxins by the method
of Eppley [17]. The results indicated that in the district of Karamoja, 44% of
105 food samples assayed contained aflatoxins and almost half of the contaminated foods contained more than 100 μg total aflatoxins/kg. In other districts of
Uganda, where hepatoma incidence was lower, both the frequency of aflatoxin
contamination (percentage of samples contaminated) and the magnitude (concentration of aflatoxins) of the contamination were also lower. From these data
the authors estimated that the daily aflatoxin (total) consumption in Karamoja
could be on the order of 20 μg to 2 mg; in comparison to studies in Kenya and
Thailand, discussed later, this exposure would be extremely high.
    Swaziland was the site of another study of aflatoxin contamination of food-
stuffs and human liver cancer [22]. Ninety cases of primary liver cancer had
been recorded from 1964 to 1968 in a newly established cancer registry; no
measure of accuracy of diagnosis or completeness of registration was given.
A sex ratio of 5:1 was observed, with males at the greater risk. Shangaan
immigrants in Swaziland in the age group of 25 to 64 had a liver cancer
incidence of 45.9/100,000/year compared to Swazis of the same age with an
incidence of 25.9/100,000/year. Samples of peanuts were collected from
various parts of the country and analyzed for aflatoxins. Frequency of contamination was high in those areas in which there was high frequency of primary
liver cancer. In addition it was determined by interview that the Shangaans

consumed larger amounts of peanuts more frequently than did the Swazis, thus supporting a possible association between aflatoxins and human liver cancer.

Prior to 1967, field studies on aflatoxin exposure did not measure toxin consumption, only contamination of relatively small numbers of food samples; also, the incidence of liver cancer was not measured directly but relied on cancer registries. In 1967 a 3-year study in Thailand began, with the intent of measuring directly both the aflatoxin consumption of carefully defined populations and the incidence of hepatocellular carcinoma in those same populations. This study began with a preliminary 23-month investigation of contamination of foods and foodstuffs by toxigenic molds [43] and aflatoxins [44] in order to establish a basis for an epidemiological study. The object of the food survey was to determine (a) the extent of mold invasion in Thai foodstuffs, (b) the commodities most often contaminated with molds and with aflatoxins, (c) the geographic and seasonal distributions of these contaminations, and (d) the capacity of the invading molds to produce toxins other than aflatoxins. Over 2000 samples, representing more than 170 kinds of foods and foodstuffs, were collected from markets, storehouses, distributors, farms, and homes. Many of the molds, isolated from food samples, were capable of producing substances other than aflatoxins which were lethal to rats, but at that time analytical methods were not available to permit measurement of the extent of contamination in the market samples; thus, aware that other mycotoxins were present in the diet, the study proceeded to the epidemiological phase based solely on aflatoxin consumption.

Having determined the geographical and seasonal distribution of aflatoxins in foodstuffs, 144 households in nine villages were randomly selected in three areas of Thailand where toxin contamination of market samples was high (Singburi), intermediate (Ratburi), and low (Songkhla). Aflatoxin consumption through cooked foods was determined by three 2-day surveys over a 12-month period, one survey during each of the three seasons [45]. In the area of Thailand in which aflatoxin contamination of market samples was high (Singburi), the daily consumption of aflatoxin B_1 was 51 to 55 ng/kg body weight on a family basis, i.e., calculated for the family as a whole rather than calculated for each individual member of the family. Daily consumption of all aflatoxins in this area was 73 to 81 ng/kg. In the area of intermediate contamination of market samples (Ratburi), daily consumption was 31 to 48 mg B_1/kg and 45 to 77 ng total aflatoxins/kg. In the south (Songkhla) where market samples contained the least toxin, daily consumption was approximately 5 ng/kg, mostly aflatoxin B_1. Variation was great between members within the families, between families in the same villages, and between villages in the same area, but the pattern was clear and is reflected in the means calculated over the period of a year. Maximum daily aflatoxin consumptions for members of surveyed households were as high as 4251 ng B_1 (6541 ng total)/kg body weight (family basis) in central Thailand (Singburi), 1284 ng B_1/kg (1701 ng total/kg) in Ratburi, and 57 ng B_1/kg (114 ng total/kg) in Songkhla.

The market survey indicated that peanuts, dried corn, and millet were the principal sources of aflatoxin in Thai foodstuffs, and that rice, the staple of the Thai diet, was relatively free of the toxin. The dietary survey indicated that

corn and millet were not often eaten and thus not important toxin sources in the
diet, but rice was important; the dietary survey suggested that leftover cooked
foods, especially rice, were a significant source of dietary aflatoxins (this
observation, of course, would be missed in a market survey). The market
survey did indicate, however, that garlic, dried chili peppers, and dried fish
were appreciable sources of the mycotoxin; this was confirmed by the dietary
survey insofar as cooked foods in which these were ingredients often contained
aflatoxins.

At the same time the dietary survey was being conducted, the incidence of
primary liver cancer was being measured directly in comparable populations
[46]. Age-specific death rates for hepatocellular carcinoma among persons
15 years of age or older were determined in Ratburi and Songkhla by investi-
gating all deaths in hospitals and homes within prescribed boundaries for the
12-month period. Liver viscerotomy specimens were collected wherever
possible for histopathological verification of the diagnosis. Six cases of
primary liver cancer were found in the Ratburi population of 97,867, supporting
the hypothesis that aflatoxin consumption is related to primary liver cancer in
Thailand.

Peers and Linsell [34] carried out an extensive study on dietary aflatoxins
and human liver cancer in Kenya. Dietary customs were such that by sampling
the evening meal one included all the dietary components used by the people and
could calculate the daily aflatoxin consumption. Samples were collected from
eight cookpots each in 16 cluster centers throughout the Murang'a district four
times a year for 21 months to permit a measure of seasonal and annual variation in aflatoxin consumption. From data obtained by the Kenya Cancer
Registry for the period 1967 to 1970 incidence rates of primary liver cancer for
populations of three different areas of high, middle, and low altitude were
calculated. The mean daily total aflatoxin intakes ranged from 3.46 ng/kg body
weight for women living at the high elevations to 14.81 ng/kg body weight for
men at low elevations. Relating these data to incidence rates in the same populations yielded a regression line $y = 19.06 \log x - 10.16$, for a correlation
coefficient of 0.87 for 4 degrees of freedom ($0.05 > P > 0.02$). This study
offers strong support for aflatoxin as a causative factor in the etiology of human
liver cancer in Kenya; and this support is further strengthened by the fact that
the data obtained from the Thailand study fit well on the Kenya regression line.

The conclusion that can be drawn from these studies in Thailand and Kenya
is that there is reasonably strong epidemiological evidence incriminating aflatoxin B_1 as having an active part in hepatocarcinogenesis in man, at least in
Africa and Southeast Asia. The argument that these levels of exposure are
barely detectable and therefore are unlikely to evoke a response is met by the
incredible potency of aflatoxin B_1 to induce tumors; as little as 1 μg/kg in the
diet produces liver tumors in rats [50], and one-tenth this concentration will
produce liver tumors in trout [20]; thus the aflatoxin intakes measured in
Thailand and Kenya are comparable to carcinogenic levels for the rat and trout.
Specifically, a 250-g rat with a daily intake of 15 g of food containing 1 μg
aflatoxin B_1/kg diet is consuming 60 ng B_1/kg body weight/day and the tumor
rate will be 10,000/100,000 rats over their life span. In Thailand and Kenya,

the human consumption of aflatoxin B_1 was up to approximately 50 ng/kg body weight/day with a tumor incidence of 6/100,000 people per year, not per life span. Moreover, the human studies in reality measured average doses, when in fact individual doses may well have exceeded these levels.

Cancer of the liver may not be the only consequence of man's chronic exposure to aflatoxins; some evidence, albeit less convincing than that which exists for the cancer relationships, suggests these mycotoxins may be a factor in a disease known as Indian childhood cirrhosis. This widely occurring disease, with an incidence peak at 3 years, is characterized by fatty infiltration of liver cells with degeneration, fibrosis, and hepatomegaly and in advanced stages proceeds to jaundice, ascites, and hepatic coma [39]. Robinson [36] explored the possibility that aflatoxins could be involved in this disease by examining human breast milk and urine for the presence of the toxins. From 43 samples of breast milk from mothers of cirrhotic children, three contained a compound with thin-layer chromatographic properties similar to those of aflatoxin B_1, whereas milk from mothers of healthy children was negative. Analysis of 43 urine specimens from cirrhotic children indicated 18 contained an aflatoxin B_1-like compound; 8 of 17 urine specimens from healthy children also contained similar material. Amla and co-workers [5] also studied breast milk and urine and obtained similar results. Yadgiri and associates [53] failed to detect aflatoxinlike material in urine from cirrhotic children but unfortunately collected the urine over HCl which rapidly decomposes aflatoxin B_1.

As a protein supplement to Indian children suffering from kwashiorkor, a peanut meal was added to the diet providing approximately 30 to 60 g of supplement per day per child [6]. Twenty children between the ages of 1.5 and 5 years consumed the supplemented diet for 5 to 30 days before it was discovered that a sample of the peanut meal contained 300 μg aflatoxins/kg; 18 of the children were being treated for kwashiorkor, one for nephrosis, and one was normal. At the time the aflatoxin contamination was discovered, 16 of the children had developed soft hepatomegaly, which became firm 2 months after toxin exposure ceased. Liver biopsies taken 1 to 2 months after peanut meal withdrawal revealed moderately severe fatty infiltration and focal necrosis; 4 months after withdrawal perilobular fibrosis was evident. By the 10th month there was complete loss of lobular architecture with dissection of septa, focal necrosis, and bile duct proliferation. These lesions, which are not characteristic of kwashiorkor, are reported to be identical to those seen in cases of Indian childhood cirrhosis [6].

3.9.3.2   Acute Toxicity of Aflatoxin B_1

If the contamination of man's food supply by the aflatoxins can reach the extent to which it can induce tumors in certain populations, then it follows that in those same populations there may develop occasional situations in which a single or short-term exposure to a high concentration of toxin is sufficient to precipitate an acute response. Reports of possible acute aflatoxin poisonings in humans have come from Taiwan [25], Uganda [40], and Thailand [12] where the incidence of liver cancer is high.

In March of 1967 26 persons in two Taiwan farming villages became ill
from what appeared to be an intoxication [25]. The seven adult cases com-
plained only of a mild, general malaise, but the 19 affected children suffered
edema of the lower extremities, abdominal pain, vomiting, and palpable liver
but no fever; three of the children aged 4 to 6 years died 6.5 h to several days
after onset of illness.

The victims were members of three families living in 10 households; two
of the families were within a 30-min walk of each other but the third family was
5 km (2 h) away. One family of five households had 14 cases, all living in two
of the households and involving every member. Whole rice from the affected
households was black-green and moldy, whereas rice from the three unaffected
households was of higher quality. The moldy rice was eaten for about 2 weeks
before the onset of illness and the death of an 8-year-old boy; aflatoxins were
not detected in samples of this rice. The second family of four households had
10 cases involving two of the households; again, one household of nine people
became ill except for a 2-month-old infant fed breast milk only. The affected
households raised and harvested their own rice, which was black and moldy and
had been eaten for about 3 weeks before the illness. This rice contained
approximately 200 μg aflatoxin B_1/kg. One child of 5 years died after being ill
for 6.5 h. Family members of other households but living with members of
affected households ate rice from an outside source and did not become ill. A
third family living in one household had two of its four children become ill after
eating moldy rice; a 4-year-old child died 3.5 days after onset of illness;
analysis of a rice sample failed to indicate the presence of aflatoxins. Unfortunately, postmortem examinations could not be done and the histopathological
changes remain unknown; thus aflatoxin B_1 can only be suspected of being
causally related to the illness.

Another suspected aflatoxin poisoning occurred in Uganda 3 months after
the Taiwan outbreak [40]. A 15-year-old boy developed abdominal pain and
swelling and 4 days later was admitted to hospital with edema of both legs, a
palpable, tender liver, and no fever, symptoms similar to the Taiwan cases.
He was diagnosed as suffering from heart failure and treated with digoxin and
the mercurial diuretic, mersalyl. His condition continued to deteriorate and
he died 2 days after admission. An autopsy revealed pulmonary edema, flabby
heart, and liver necrosis; a histopathological examination indicated interstitial
edema of the heart, congestion and edema of the lungs, and mild fatty change
and centrilobular necrosis of the liver.

The youth's 6-year-old sister and 3-year-old brother became ill at the
same time with abdominal pain and malaise. The brother recovered without
treatment and the sister recovered after 6 days of treatment at a local dispens-
ary. The family diet consisted mainly of cassava and also beans, fish, and
meat. The cassava in the home was moldy and contained 1.7 mg aflatoxins/kg.
Based on data obtained from studies on aflatoxin and the young African monkey
(Cereopithecus aethiops) [1], this level of toxin probably could have produced a
fatal response after eating the cassava for a few weeks.

More direct, although still circumstantial evidence involving aflatoxin in
acute poisonings in man comes from Thailand [12, 42]. Reye's syndrome [35]
is epidemic in northreastern Thailand [11] and is limited to children up to

adolescence [10, 33]. It is characterized by a short prodrome of several hours followed by vomiting, hypoglycemia, convulsions, hyperammoniemia, and coma usually ending in death 24 to 48 h after onset; histopathologic examination reveals severe cerebral edema and extensive fatty accumulation in hepatocytes, renal tubular epithelium, and myocardial fibers. A particular case of Reye's syndrome occurred in Thailand and suggested that aflatoxin might have a role in the disease [12].

A 3-year-old Thai boy ate only leftover cooked rice for 2 days. The rice had become moldy due to lack of refrigeration but remained palatable; chemical analysis indicated more than 10 mg total aflatoxin/kg. On the third day, after having been ill for 12 h with fever, vomiting, coma, and convulsions, he was admitted to hospital; his blood glucose was 24 mg/100 ml and he died 6 h after admission. An autopsy and histopathologic examination indicated Reye's syndrome.

Earlier studies on rodents, fowl, farm animals, fish, cats, dogs, rabbits, and even monkeys did not detect a response to aflatoxins which was similar to Reye's syndrome [30]. Sparse primate data prompted a detailed study of acute aflatoxicosis in this species in an effort to compare the intoxication to the children's disease [12, 42]. A standard LD_{50} determination was conducted on six groups of four young, female crab-eating macaques (Macaca fasicularis). The intoxication in the monkey was remarkably similar to Reye's syndrome. The monkeys coughed and vomited 12 to 72 h after oral administration of the toxin, went into coma with occasional convulsions; serum glucose and phospholipid levels fell and nonesterified fatty acid levels rose. Necropsy revealed cerebral edema and marked fatty degeneration of the liver, heart, and kidney. However, bile duct hyperplasia and liver cell necrosis were more extensive than seen in any of the Thai cases of Reye's syndrome. Nuclear inclusions were found in the pancreatic acinar cells, which at that time had not been reported in Reye's syndrome cases; however, in 1974 Collins [15] reported finding such inclusions in five cases of the children's disease in the United States.

The acute toxicity test produced a large number of tissue specimens which were chemically analyzed for aflatoxins [42]. Unmetabolized aflatoxin B_1 could be detected in monkey tissues up to 148 h after administration of even a sublethal dose. Bile contained up to 162 to 176 μg/kg and brain up to 30 μg/kg. This information suggested a reasonable chance of detecting aflatoxin B_1 in Reye's syndrome cases, if indeed aflatoxins were responsible in any way for the disease.

Autopsy specimens from 23 Thai children who died with Reye's syndrome and from 15 children and adolescents who died from unrelated causes were chemically assayed for aflatoxins [41]. Aflatoxins B_1 and B_2 were found in specimens from 22 of the 23 Reye's syndrome cases, usually at levels between 1 and 4 μg/kg; however, in two cases the concentrations of aflatoxin B_1 were of the same magnitude as the levels in the livers from monkeys in the acute toxicity study just discussed. Intestinal contents from six human cases contained as much as 127 μg B_1/kg and 15 μg B_2/kg. Traces of aflatoxin B_1 were detected in autopsy specimens from 11 of the 15 control subjects, probably reflecting chronic low level ingestion of the toxin by the general population of northeastern Thailand, an area in which aflatoxin contamination of the food supply is high [44].

Not all of the evidence associating aflatoxins with Reye's syndrome comes from Thailand. Becroft and Webster [9] found apparent aflatoxins B_1 and G_1 in liver extracts from two New Zealand cases of Reye's syndrome. In Czechoslovakia Dvorackova [16] has found aflatoxins in six of seven cases of Reye's syndrome and a similar disease, but characterized by more extensive liver damage resembling that seen in Indian childhood cirrhosis. In two cases of sisters the mother worked on a poultry farm during both pregnancies and was in daily contact with the fodder; cows fed the poultry fodder produced apparent aflatoxin in their milk.

Several hypotheses for the etiology of the disease have been offered and can be grouped into two classifications: Reye's syndrome may be a viral infection, or it may be an intoxication. Factors favoring a viral origin include (a) seasonal variation, (b) prodromal symptoms including fever and occasionally upper respiratory tract infections, and (c) association of a number of cases with chicken pox or influenza. Factors against this argument include (a) lack of family involvement, (b) failure to associate a particular virus with the disease [33], and (c) lack of pathological findings characteristic of known viral diseases. There is no evidence of encephalitis; indeed, fatty liver is not usually associated with encephalitis. The liver histology is not consistent with viral hepatitis. Glick and co-workers [18] reported on 62 cases seen in southern and central United States and Puerto Rico from January 1967 through June 1969; seven of these cases were associated with prodromal chicken pox and 48 had an upper respiratory, influenzalike illness. Becroft [8] pointed out that, although unlikely possibilities, an unknown virus may infect and alter the liver without causing necrosis, or that there may be present a distant infection by an organism which produces a hepatotoxin.

Some aspects of the disease are difficult to explain solely on the basis of an intoxication. In particular, these include (a) lack of sibling or other family involvement, (b) rarity of reports of "sublethal" doses, and (c) in spite of the ubiquity of toxigenic strains of Aspergillus flavus (producers of the aflatoxins; Sec. 1.2.3.3), the improbability that children in such diverse places were exposed to and could ingest lethal quantities of the poison.

It has been estimated that in Thailand young children could consume as much as 160 μg aflatoxin B_1/kg at one time from heavily contaminated but "edible" peanuts [45]; this is approximately 50 times less than the LD_{50} for the monkey [42]. As suggested by Bourgeois and co-workers [11] the pathogenesis of Reye's syndrome perhaps is initiated by a toxic, nutritional, or infectious insult, resulting in a subclinical liver injury, which may be manifested by fatty metamorphosis; a second injury to the liver, acute aflatoxicosis, could then be far more effective than if it occurred in a completely healthy individual, and hence, less aflatoxin than predictable from the monkey study could be fatal.

These reports, even in aggregate, only suggest that the aflatoxins may play a role in the etiology of Reye's syndrome, at least as it occurs in Thailand, and perhaps New Zealand and Czechoslovakia. This suggestion should be tested in the laboratory and in the field. The same geographical area in Thailand in

which Reye's syndrome occurs in high numbers is also an area in which pro-
vincial hospitals see high numbers of liver cancer patients. Another question
that should be studied in the laboratory and in the field is whether survivors of
Reye's syndrome in Thailand have a higher risk of later developing cancer of
the liver.

3.9.4   Problems in Epidemiological
          Design

3.9.4.1   Selection of Study Populations

Generally, associations between exposure to mycotoxins and human disease will
be made epidemiologically, and there are some important problems in designing studies to investigate such associations. Since the source of exposure to
mycotoxins usually will be food, study populations must be those in which
dietary surveys are practical, unless mycotoxin consumption can be adequately
determined by means other than a dietary survey, for example, quantitative
measurement of the toxin or its metabolites in blood or urine. Such survey
populations should have an uncomplicated diet; if the health problem being
studied is a chronic disease, the diet of the survey population should not have
changed appreciably over the induction period.
     Ideally, there will be only one variable between the reference and experi-
mental populations and the two groups will be comparable with respect to
genetics, age, sex, socioeconomic position, etc.; although the ideal cannot be
reached, one must work with as few variables as possible. The size of the
study population will be governed by the expected magnitude of the incidence of
the disease, the expected difference between the incidences in the reference and
experimental populations, and resources available to the investigator; the size
will be inversely proportional to both the magnitude of the incidence and the
magnitude of the difference between the two measured incidences.

3.9.4.2   Mycotoxin Occurrence and
            Consumption

Field studies on mycotoxin contamination of foods and consumption of these
compounds are expensive, but usually more information can be obtained from
the food samples than can be used at the time of the study, or other factors not
recognized at that time may subsequently prove important, and it would be
desirable to go back to the original food samples for more data. Peers and
Linsell [34] recognized this and placed in storage ($-30°C$) for future examination a complete 1 year's collection of diet samples from the Kenyan liver cancer
study discussed earlier. When such storage facilities are not available, the
technique used in the Thailand liver cancer study might be used; each food
sample was cultured and soil spore suspensions for all molds isolated from the
cultures were prepared for future reference should the need to demonstrate
their toxigenicity arise.

When a food sample is collected, the history of that sample should be recorded; these data can later be computer analyzed for trends leading to mycotoxin production in man's food supply.

Market surveys can be useful preliminary studies to define field conditions; they can define which mycotoxins to consider, what factors in the local area govern the occurrence of the toxins, which dietary items are most likely to be contaminated, and when and to what extent the contamination occurs.

Plate samples are more relevant to the measurement of toxin consumption but are much more difficult to obtain. Also, members of the dietary survey population may consume toxin via snacks and other unobserved sources, so that plate samples may give only minimum toxin consumption levels. A more satisfactory measure of toxin consumption, perhaps, could be based on levels of the toxin or one of its metabolites in blood or urine; this could give the most accurate estimate.

3.9.4.3   Laboratory Assays

Laboratory support for epidemiological field work can be critical to the success of the study. For mycotoxin problems, it is here that one obtains the quantitative data on the levels of various compounds in samples collected from the field. To be fully effective the assays should be simple so that they can be learned by personnel not scientifically trained. The procedure should be capable of routine use with instrumentation that can be maintained and repaired locally. Chemical assays should require instruments no more complicated than perhaps a spectrophotometer; assays that can be done in hotel rooms or the back of a car are always appreciated. Mycological assays can be simple and in tropical countries do not require even an incubator. Bioassays for unknown toxins or the mycotoxin in question can be troublesome but oftentimes provide results of great value. Rodents are usually used for economy. Feeding a moldy diet may result in rejection by the animal and extracts of the field samples may be needed. Extracts have the advantage of probable concentration of the bioactive agent and versatility in using various routes of administration. If the foodstuffs in the local area of the field vivarium are often naturally contaminated with mycotoxins, preparation or procurement of a chow diet may present a problem. In any event, having a laboratory located in or near the field minimizes transportation changes occurring in the biological specimens and thus laboratory results more accurately reflect what occurs in the field.

3.9.4.4   Relevant Cofactors

Defining the etiology of a human mycotoxicosis may not be as simple as relating the disease to levels of exposure to a toxic agent. The nutritional quality of the diet for the population at risk may have an important role in the cause of the disease. This may be the case in alimentary toxic aleukia in Russia where the victims often suffered from prolonged poor nutrition [27]; other examples may be Indian childhood cirrhosis in kwashiorkor cases consuming aflatoxins [6], or Reye's syndrome children in Thailand ingesting aflatoxins [11].

Dietaries contaminated with a mycotoxin might reasonably be expected to contain, at least occasionally, one or more other microbial toxins or plant toxins, which could influence the toxicity of the agent under study. It thus is important to be aware of these secondary toxins, and how often and at what levels they occur. Besides other toxins, there may be other diseases in the study population which may influence the relationship between the toxin and disease under study. Thus, the effect of liver fluke and viral hepatitis on the association between mycotoxins and liver cancer needs study.

3.9.5   Summary

The number of human mycotoxicoses, at least those which occur on a scale large enough to be detected epidemiologically, is small. That is not to say human mycotoxicoses are unimportant. In the past, the ergot alkaloids and the mycotoxin(s) responsible for alimentary toxic aleukia in Russia have caused severe human suffering. Evidence is accumulating to include the aflatoxins among those presenting a health problem.

More recently discovered mycotoxins are beginning to appear as possible threats to man's health. Glinsukon et al. [19] have associated cytochalasin E with human diet; an analog, cytochalasin B (Sec. 2.9), not yet demonstrated in foods, is a potent teratogen [26]. Wilson and associates have studied phytoalexins in mold-damaged sweet potatoes and found furanoterpenoids which are highly toxic to rodent lung [14]; since these compounds are produced not by the mold, but by the sweet potato in response to mold damage, it is quite likely that humans, even after removing the blighted portion of the tuber, ingest these toxins.

The most relevant question, and the most difficult to answer, is what is to be done about mycotoxins in the human environment. Exposure must be minimized, of course, but to what level? At what level of mycotoxin should food be condemned or destroyed? The logical approach to these questions is to find a level at which the risk to human health is tolerable. Unfortunately, a scientific means of determining these levels is not available; and who is to bear the burdensome responsibility of determining what risks to human health are tolerable? One must not quit in despair, but strive even more diligently, in the laboratory with animals and in the field with humans, to seek better answers to the problem of mycotoxins and human health.

References

1. E. Alpert and A. Serck-Hanssen (1970): Aflatoxin-induced hepatic injury in the African monkey. Arch. Environ. Health 20:723-728.
2. M. E. Alpert, M. S. R. Hutt, and C. Davidson (1968): Hepatoma in Uganda: A study in geographic pathology. Lancet 1:1265-1267.
3. M. E. Alpert, M. S. R. Hutt, and C. Davidson (1969): Primary hepatoma in Uganda: A prospective clinical and epidemiologic study of forty-six patients. Am. J. Med. 46:794-802.

4. M. E. Alpert, M. S. R. Hutt, G. N. Wogan, and C. S. Davidson (1971): Association between aflatoxin content of food and hepatoma frequency in Uganda. Cancer 28:253-260.

5. I. Amla, S. Kumari, V. Sreenivasmurthy, P. Jayaraj, and H. A. B. Parpia (1970): Role of aflatoxin in Indian childhood cirrhosis. Indian Pediatr. 7:262-270.

6. I. Amla, C. S. Kamala, G. S. Gopalakrishna, A. P. Jayaraj, V. Sreenivasamurthy, and H. A. B. Parpia (1971): Cirrhosis in children from peanut meal contaminated by aflatoxin. Am. J. Clin. Nutr. 24:609-614.

7. J. R. Bamburg, F. M. Strong, and E. B. Smalley (1969): Toxins from moldy cereals. J. Agric. Food Chem. 17:443-450.

8. D. M. O. Becroft (1968): Encephalopathy and fatty degeneration of the viscera. Am. J. Dis. Child. 115:750.

9. D. M. O. Becroft and D. R. Webster (1972): Aflatoxins and Reye's disease. Br. Med. J. 4:117.

10. C. Bourgeois, N. Keschamras, D. S. Comer, S. Harikul, H. Evans, L. Olson, T. Smith, and M. R. Beck (1969): Udorn encephalopathy: Fatal cerebral edema and fatty degeneration of the viscera in Thai children. J. Med. Assoc. Thailand 52:553-565.

11. C. Bourgeois, L. Olson, D. Comer, H. Evans, N. Keschamras, R. Cotton, R. Grossman, and T. Smith (1971): Encephalopathy and fatty degeneration of the viscera: A clinicopathologic analysis of 40 cases. Am. J. Clin. Pathol. 56:558-571.

12. C. H. Bourgeois, R. C. Shank, R. A. Grossman, D. O. Johnsen, W. L. Wooding, and P. Chandavimol (1971): Acute aflatoxin B_1 toxicity in the macaque and its similarities to Reye's syndrome. Lab. Invest. 24:206-216.

13. F. J. Bové (1970): The Story of Ergot. Karger, Basel.

14. M. R. Boyd, L. T. Burka, T. M. Harris, and B. J. Wilson (1974): Lungtoxic furanoterpenoids produced by sweet potatoes (Ipomoea batatas) following microbial infection. Biochim. Biophys. Acta 337:184-195.

15. D. N. Collins (1974): Ultrastructural study of intranuclear inclusions in the exocrine pancreas in Reye's syndrome. Lab. Invest. 30:333-340.

16. I. Dvorackova (1974): Personal communication.

17. R. M. Eppley (1966): A versatile procedure for the assay and preparatory separation of aflatoxin from peanut products. J. Assoc. Offic. Anal. Chem. 49:1218-1223.

18. T. H. Glick, W. H. Likosky, L. P. Levitt, H. Mellin, and D. W. Reynolds (1970): Reye's syndrome: An epidemiological approach. Pediatrics 46:371-377.

19. T. Glinsukon, S. S. Yuan, R. Wightman, Y. Kitaura, G. Büchi, R. C. Shank, G. N. Wogan, and C. M. Christensen (1974): Isolation and purification of cytochalasin E and two tremorgens from Aspergillus clavatus. Plant Foods Man 1:113-119.

20. J. E. Halver (1969): Aflatoxicosis and trout hepatoma, in L. A. Goldblatt (ed.): Aflatoxin Scientific Background, Control and Implications. Academic Press, New York, pp. 265-306.

21. A. Z. Joffe (1971): Alimentary toxic aleukia, in S. Kadis, A. Ciegler, and S. J. Ajl (eds.): Microbial Toxins, Vol. 7: Algal and Fungal Toxins. Academic Press, New York, pp. 139-189.

22. P. Keen and P. Martin (1971): Is aflatoxin carcinogenic in man? The evidence in Swaziland. Trop. Geogr. Med. 23:44-53.

23. H. F. Kraybill and M. B. Shimkin (1964): Carcinogenesis related to foods contaminated by processing and fungal metabolites, in A. Haddow and S. Weinhouse (eds.): Advances in Cancer Research, Vol. 8. Academic Press, New York, pp. 191-248.

24. E. LeBreton, C. Frayssinet, and J. Boy (1962): Sur L'apparition d'hepatomes "spontanes" chez le rat wistar. Role de la toxine de l'Aspergillus flavus. Interet en pathologie humaine et cancerologie experimentale. C. R. Acad. Sci. (Paris) 255:784-786.

25. K.-H. Ling, J.-J. Wang, R. Wu, T.-C. Tung, C.-K. Lin, S.-S. Lin, and T.-M. Lin (1967): Intoxication possibly caused by aflatoxin B_1 in the moldy rice in Shuang-Chih Township. J. Formosan Med. Assoc. 66:517-525.

26. G. P. Linville and T. H. Shepard (1972): Neural tube closure defects caused by cytochalasin B. Nature (New Biol.) 236:246-247.

27. C. F. Mayer (1953): Endemic panmyelotoxicosis in the Russian grain belt. I. The clinical aspects of alimentary toxic aleukia (ATA): A comprehensive review. Milit. Surg. 113:173-189.

28. C. F. Mayer (1953): Endemic panmyelotoxicosis in the Russian grain belt. II. The botany, phytopathology, and toxicology of Russian cereal food. Milit. Surg. 113:295-315.

29. C. J. Mirocha and S. Pathre (1973): Identification of the toxic principle in a sample of poaefusarin. Appl. Microbiol. 26:719-724.

30. P. M. Newberne and W. H. Butler (1969): Acute and chronic effects on aflatoxin on the liver of domestic and laboratory animals: A review. Cancer Res. 29:236-250.

31. A. G. Oettlé (1964): Cancer in Africa, especially in regions south of the Sahara. J. Natl. Cancer Inst. 33:383-439.

32. L. E. Olifson (1957): Bull. Chkalov Sect. D.E. Mendeliev All Soviet Union Chem. Assoc. 7:7, 37. Cited by A. Z. Joffe (1971): Alimentary toxic aleukia, in S. Kadis, A. Ciegler, and S. J. Ajl (eds.): Microbial Toxins, Vol. 7: Algal and Fungal Toxins. Academic Press, New York, pp. 139-189.

33. L. C. Olson, C. H. Bourgeois, R. B. Cotton, S. Harikul, R. A. Grossman, and T. J. Smith (1971): Encephalopathy and fatty degeneration of the viscera in northeastern Thailand. Pediatrics 47:707-716.

34. F. G. Peers and C. A. Linsell (1973): Dietary aflatoxins and liver cancer. A population-based study in Kenya. Br. J. Cancer 27:473-484.

35. R. D. K. Reye, G. Morgan, and J. Baral (1963): Encephalopathy and fatty degeneration of the viscera: A disease entity in childhood. Lancet 2:749-752.

36. P. Robinson (1967): Infantile cirrhosis of the liver in India. With special reference to probable aflatoxin etiology. Clin. Pediatr. 6:57-62.

37. M. Saito, M. Enomoto, and T. Tatsuno (1969): Radiomimetic biological properties of the new scirpene metabolites of Fusarium nivale. Gann 60: 599–603.

38. M. Saito, M. Enomoto, T. Tatsuno, and K. Uraguchi (1971): Yellowed rice toxins. Luteoskyrin and related compounds, chlorine-containing compounds, and citrinin. Citreoviridin, in A. Ciegler, S. Kadis, and S. J. Ajl (eds.): Microbial Toxins, Vol. 6: Fungal Toxins. Academic Press, New York, pp. 299–380.

39. B. C. Sen (1887): Enlargement of liver in children. Indian Med. Gaz. 22: 338; Cited by P. Robinson (1967). Clin. Pediatr. 6:57–62.

40. A. Serck-Hanssen (1970): Aflatoxin-induced fatal hepatitis? A case report from Uganda. Arch. Environ. Health 20:729–731.

41. R. C. Shank, C. H. Bourgeois, N. Keschamras, and P. Chandavimol (1971): Aflatoxins in autopsy specimens from Thai children with an acute disease of unknown etiology. Food Cosmetol. Toxicol. 9:501–507.

42. R. C. Shank, D. O. Johnsen, P. Tanticharoenyos, W. L. Wooding, and C. H. Bourgeois (1971): Acute toxicity of aflatoxin B_1 in the macaque monkey. Toxicol. Appl. Pharmacol. 20:227–231.

43. R. C. Shank, G. N. Wogan, and J. B. Gibson (1972): Dietary aflatoxins and human liver cancer. I. Toxigenic molds in foods and foodstuffs of tropical south-east Asia. Food Cosmetol. Toxicol. 10:51–60.

44. R. C. Shank, G. N. Wogan, J. B. Gibson, and A. Nondasuta (1972): Dietary aflatoxins and human liver cancer. II. Aflatoxins in market foods and foodstuffs in Thailand and Hong Kong. Food Cosmetol. Toxicol. 10: 61–69.

45. R. C. Shank, J. E. Gordon, G. N. Wogan, A. Nondasuta, and B. Subhamani (1972): Dietary aflatoxins and human liver cancer. III. Field survey of rural Thai families for ingested aflatoxins. Food Cosmetol. Toxicol. 10:71–84.

46. R. C. Shank, N. Bhamarapravati, J. E. Gordon, and G. N. Wogan (1972): Dietary aflatoxins and human liver cancer. IV. Incidence of primary liver cancer in two municipal populations of Thailand. Food Cosmetol. Toxicol. 10:171–179.

47. T. Tatsuno (1968): Toxicologic research on substances from Fusarium nivale. Cancer Res. 28:2393–2396.

48. Y. Ueno, N. Sato, K. Ishi, K. Sakai, H. Tsunoda, and M. Enomoto (1973): Biological and chemical detection of trichothecene mycotoxins of Fusarium species. Appl. Microbiol. 25:699–704.

49. K. Uraguchi (1971): Neurotoxic mycotoxins of Penicillium citreo-viride Biourge, in H. Raskova (ed.): Pharmacology and Toxicology of Naturally Occurring Toxins. Academic Press, New York, pp. 143–174.

50. G. N. Wogan, S. Paglialunga, and P. M. Newberne (1974): Carcinogenic effects of low dietary levels of aflatoxin B_1 in rats. Food Cosmetol. Toxicol. 12: 681–685.

51. R. D. Wyatt, B. A. Weeks, P. B. Hamilton, and H. R. Burmeister (1972): Severe oral lesions in chickens caused by ingestion of dietary fusariotoxin T-2. Appl. Microbiol. 24:251–257.

52. R. D. Wyatt, W. M. Colwell, P. B. Hamilton, and H. R. Burmeister
     (1973): Neural disturbances in chickens caused by dietary T-2 toxin.
     Appl. Microbiol. 26:757-761.
53. B. Yadgiri, V. Reddy, P. G. Tulpule, S. G. Srikantia, and C. Gopalan
     (1970): Aflatoxin and Indian childhood cirrhosis. Am. J. Clin. Nutr. 23:
     94-98.
54. D. Gröger (1972): Ergot, in S. Kadis, A. Ciegler, and S. J. Ajl (eds.):
     Microbial Toxins, Vol. 8, Academic Press, New York, pp. 321-373.

Ronald C. Shank

3.10 Fusarium Poae and F. Sporotrichioides as Principal Causal Agents of Alimentary Toxic Aleukia

3.10.1 Introduction

The studies described in this section constitute a comprehensive investigation of the causes and problems of alimentary toxic aleukia (ATA), a disease widespread among the population of the Soviet Union, especially the Oreburg district, during World War II and the postwar years until 1947.

In the years 1942 to 1947 ATA, a very serious and in most cases fatal disease, accompanied by extreme leukopenia, multiple hemorrhages, agranulocytosis, necrotic angina, sepsis, and exhaustion of the bone marrow, occurred widely in some republics and districts. As indicated by the author [64, 68-73] the causal agents of this disease were found to be species of fungi developing on grain overwintered in the field and exposed to the extreme conditions of winter-spring seasons prevalent in some parts of the U.S.S.R. Among the fungi found in the overwintered grains, species of Fusarium poae (Sec. 1.4.6.1) and F. sporotrichioides (Sec. 1.4.6.2) were most frequently encountered [64, 68, 69].

The disease appeared in 1942, increased considerably in 1943, and especially in 1944, when the food situation continued to deteriorate, and large parts of the population were reduced to collecting grain from fields that had been covered by snow throughout the winter-spring months.

Earlier researches on the conditions attending outbreaks of ATA, the toxicity of overwintered cereals, the effects of toxins on animals, and the phytotoxic effects on plants have been reviewed by Joffe [70-75].

In this study, we describe the epidemiological background and etiology of the disease, bioassay methods and procedures with personal analysis, toxic fungi, and their toxins, taxonomic problems of Fusarium of the section Sporotrichiella, general clinical syndromes, and organs affected and their pathology.

3.10.2 Epidemiological Background

The disease of septic angina or alimentary toxic aleukia (ATA) has been recorded in Russia at various times, probably as early as the nineteenth century. In 1913 this disease was observed again in Eastern Siberia and in the Amur region [206, 207]. After three decades the areas of outbreaks widened and in the spring of 1932 reappeared suddenly in an endemic form in several districts of Western Siberia [27, 42, 44, 82] but did not receive the serious attention it deserved.

The symptoms of the disease of ATA were fully described in the Russian literature and, much later, incompletely in some American articles [41, 114, 115].

The typical symptoms of the disease include hemorrhagic rash on the skin, leukopenia, agranulocytosis, bleeding from the nose, throat, gums, and genital tract, necrotic angina, hemorrhagic diathesis, sepsis, and exhaustion of the bone marrow [25-28, 101, 105, 107-109, 133, 135, 154, 206, 207].

3.10.2.1   Review of Outbreaks of ATA

In May and June of 1934, the disease appeared in Western Siberia [192]. After a few years ATA reoccurred in Ryazan, Molotov, Sverdlovsk, Omsk, Novosibirsk [131], Altai territory [58, 168, 169, 206], and some counties of Kazakhstan and Kirghiz S.S.R. [96, 97]. The disease became widespread in 1942 and, according to the data presented in the report of Beletskij [16], appeared again at the beginning of World War II in different republic and districts, including Molotov, Kirov [137], Saratov [181], Gorkov, Yaroslavsk, Kuybyshev, Chelabinsk, Orenburg of the Ural [5, 27, 28, 100, 108, 109, 117, 154], and also in the Udmurt A.S.S.R. [182], as well as in the Tartar A.S.S.R. [34, 35, 101] and Bashkir A.S.S.R. [55, 193, 194, 180, 184, 211]. In 1943 the disease appeared in the Leningrad [134], Uljanovsk [194], and Stalingrad districts and also in the Moldavian S.S.R. and Mari A.S.S.R. [16]. In 1944 the disease spread considerably and appeared in the northwest and southeast sections of the Ural [16, 27, 28, 45, 167], and approached regions of the Volga River [16]. It occurred in the Ukrainian S.S.R., in central regions of the European Soviet, and also in Central Asia and in Far Eastern Siberia [163].

All in all, in 1944 there were outbreaks of ATA in 34 districts and counties. In that year the food situation deteriorated further and much of the population was reduced to collecting grains that had been left in the fields throughout the winter months. In 1945 the disease occurred in the Voronezh district [135], in Komi A.S.S.R., and in 12 other districts and regions [163]. In 1946, ATA was observed for the first time in the Kostromsk district, in Kabardino-Balkarsk A.S.S.R., and in other areas, for a total of 19 districts and counties [164]. In 1947, ATA appeared in 23 regions, such as Tomsk, Omsk, Arkhangelsk, and Novosibirsk, among others. The frequency of occurrence of the disease in the U.S.S.R. was found to be highest in latitudes of 50 to 60° and longitudes of 40 to 140° [115]. In this zone the disease of ATA recurred several times [169]. For example, in the Altai territory, the outbreaks were described every year for 14 years [206, 207]; in the Molotov district, for 8 years; in Bashkir A.S.S.R., 7 years [193, 194, 211]; and in 6 out of 31 regions affected by ATA, it appeared only once, e.g., in Komi A.S.S.R. [163].

The epidemiological investigations proved that ATA occurred in families which gathered various grains from the fields in the spring after the snow thawed. In certain sporadic cases, the disease occurred in different seasons

because people purchased the overwintered toxic grains, meal, or products
prepared from this grain in the markets or in the affected villages. In 1945 to
1947 the disease occurred comparatively seldom and was mainly caused by
instances in which overwintered toxic cereal crops were sold.

The mortality in 1942 to 1944 was high, and whole families, or even entire
villages, were affected, mostly in agricultural areas.

Research into the nature and cause of the condition was carried out
initially by different investigators simultaneously without any central planning
or coordination. The fact that the typical clinical syndrome of ATA could not
be reproduced in laboratory animals also hindered research. In 1932 the
nature of the disease was still unknown, and outbreaks were erroneously
labeled as diphtheria or cholera.

In view of the sudden outbreaks and high mortality (up to 60%) ATA was
considered to be an epidemic disease of infectious origin; however, epidemiological and bacteriological studies did not confirm this hypothesis. Moreover,
the fact that none of the medical staff who took care of the patients was ever
affected by ATA led to rejection of the epidemic hypothesis.

For quite a long period ATA was considered to be due to deficiency of
vitamins B_1 and C and riboflavin. This hypothesis, like others relating ATA to
bacterial infection, was rejected. Although a deficiency of riboflavin was
detected in experimental animals suffering from severe aleukia, agranulocytosis, and anemia, the diet of patients with ATA was not deficient in this
vitamin [81, 151, 159].

Belief in these theories delayed recognition of the true nature of ATA.
Eventually, however, it was realized that the disease was caused by the
ingestion of snow-covered overwintered grains, or their products, which had
formed the staple diet of the peasant population in the agricultural areas in
Russia and which had been infected by toxic fungi [71-73].

The most important question at this stage was to determine which specific
fungi were responsible for the intoxication. Eventual identification of the cause
of the disease led to the introduction of proper prophylactic measures.

The studies described here deal especially with the Orenburg district where
the author of this chapter worked for 8 years for the Institute of Epidemiology
and Microbiology of the U.S.S.R. Ministry of Health and headed the Mycological
Division of the Institute. This Institute established a special laboratory of ATA
(septic angina) for the investigation of all aspects of the disease. The following
were investigated: the role of overwintered cereal crops, the mycoflora of these
grains, the toxic properties of cryophilic fungi developing at especially low
temperatures, climatic and ecological conditions for toxin production in grains,
characteristics of these toxins in animals, the chemistry of toxins, and,
mainly, the clinical symptoms and pathological findings in man.

The first outbreaks in the Orenburg district were in 1924 and in 1934 [45].
In 1942 and 1943 the disease reappeared on a large scale. In the spring of 1944
it reached colossal dimensions, affecting 47 out of 50 counties. The Orenburg
district has a conterminal boundary with the Kuybyshev, Saratov, and Chelabinsk districts, Tatar and Bashkir A.S.S.R., and also with Kazakhstan S.S.R..
In 1942 ATA appeared in 15 counties in the northwest and in four counties in the

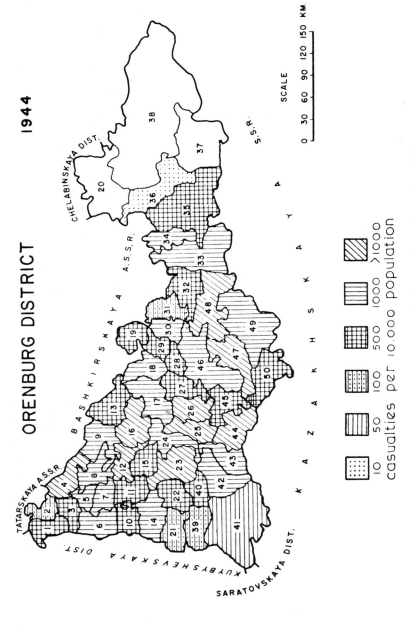

Fig. 1.   The peak year of outbreaks in various counties of the Orenburg district in 1944.

central portions of the district. In the spring of 1943, the disease affected 30 out of 50 counties in the north, west, and south of the district. The outbreaks peaked in 1944, when 47 out of 50 counties were affected (only three easterly counties were not affected). In that year the population in the Orenburg and other districts of Soviet Russia suffered enormous casualties. It can be seen from Fig. 1 that more than 10% of the population was affected, and many fatalities occurred in 9 of the 50 counties of the district [71, 72, 75]. In 1945, 1946, and 1947 the disease declined and affected 14, 8, and 12 counties respectively. In 1948 and 1949 the disease was absent.

3.10.2.2   Factors Affecting Outbreaks
           of ATA

The occurrence of ATA in man was found to be affected by the factors considered in the following sections.

   3.10.2.2.1   Quantity of Toxic Food Ingested. ATA usually developed after the consumption of at least 2 kg of food prepared from toxic overwintered grain. Lesions in the hematopoietic system were produced as a result of accumulation of toxic material in the body. Death occurred after eating 6 kg or more [28, 108, 109, 117, 135, 153, 207].

   3.10.2.2.2   The Kind of Cereal Ingested. Prosomillet and wheat are the most toxic cereals when overwintered in the field.

   3.10.2.2.3   Time Lag in Development of Symptoms. Symptoms usually appeared 2 to 3 weeks after consumption of toxic grains (prosomillet, wheat, barley, rye, oats, buckwheat). For example, when a patient consumed toxic grain in sufficient quantity at the end of April, he became sick with the disease of ATA in the middle of May. Death occurred 6 to 8 weeks after the initial eating of the toxic grain.

   3.10.2.2.4   Concentration of Toxin in the Food. Great variations were found in the toxicity of overwintered grains, even within samples of one field. Entire families were found who had ingested highly toxic grains and were all affected by ATA. Families who fed on nontoxic overwintered grain were not affected.

   3.10.2.2.5   Sensitivity of the Individual. The disease of ATA may develop in anyone who consumed food that had been prepared from toxic over- wintered grains.

   3.10.2.2.6   Age. Breast-fed babies less than 1 year old were not affected since the toxic substance was not secreted in the milk of the sick mother [28, 108, 174, 176]. The disease occurred in infants of 1 year of age and older if they had eaten products of toxic overwintered cereals [101, 174- 177, 207]. Disease occurred most frequently and caused the highest mortality between the ages of 8 and 50 years [28, 154].

3.10.2.2.7   Sex. The literature did not mention any sex differences in the mortality of ATA. We also suggested that there was no difference in the incidence of the disease in men and women, although Talayev et al. [192] noticed a preference of the disease for middle-aged women.

3.10.2.2.8   Nutritional Status. Populations fed on balanced diets were less sensitive to the toxin than undernourished persons. Those whose diet consisted almost entirely of overwintered cereals were more severely affected.

3.10.2.2.9   Environmental and Agricultural Factors. The annual variation of ATA outbreaks depended on meteorological and agricultural factors.

1. The Season of Harvesting. Grains harvested during the spring thaws were toxic; those harvested in autumn were not. Toxin formation was particularly pronounced in spring.
2. Weather. Weather factors favoring toxin formation in late winter and chiefly spring were extraordinarily deep and plentiful snow cover, which prevented the soil from freezing to its usual depth, and relatively high temperatures leading to frequent thawing and freezing. (The mean temperature of the months January-February 1944 was only -8.3°C, as compared with the much lower 10-year average of -14.7°C for these months.) Light autumn rainfall may also be a contributory factor to ATA outbreaks [71].
   The combination of much moisture and moderate temperatures permitted abundant growth of various soil fungi on the grain. The biological activity of these fungi produced different degrees of toxicity, especially species of <u>Fusarium</u> (Sec. 1.4.6) which is most highly toxic. The toxic ingredients appeared about as frequently in the soil as in the cereal grains, but vegetative parts of the plants were less frequently toxic [71-73].
3. Altitude. The disease was not found in the Orenburg district at altitudes higher than 350 to 400 m above sea level probably because of their lower winter temperatures.
   In 1944, the year with the highest number of cases, in which 47 counties were affected (see Fig. 1), not a single case was recorded in the three most easterly counties despite the fact that the population of this area consumed overwintered grains to the same extent as in other counties. All the other counties, in which the number of cases was moderate to very numerous, were at an altitude of 70 to 200 m above sea level [71]. Yefremov [207] also stated that in low areas overwintered grains were more toxic.

3.10.3   Bioassay Methods and
         Procedures

The toxic grain responsible for outbreaks of ATA was investigated chiefly by skin tests on rabbits. This skin test is, at present, a generally accepted laboratory test for toxicity of overwintered cereal grains and also for toxins produced by various fungi, especially by <u>Fusarium</u> of the <u>Sporotrichiella</u> section, developed on overwintered grains in the field which cause ATA.

3.10.3.1   Toxicity Test on Rabbits

The mycotoxins of <u>Fusarium poae</u> (Sec. 1.4.6.1), <u>F. sporotrichioides</u> (Sec. 1.4.6.2), <u>F. sporotrichioides</u> var. <u>tricinctum</u> (Sec. 1.4.6.3), and other isolates were assessed by skin tests performed on male and female rabbits weighing 1.5 to 2 kg. Only rabbits with nonpigmented skin were used because the skin was thinner and they were more sensitive to toxins [68, 69, 73].

Each rabbit was kept in a separate cage, and was maintained on laboratory feed. On the back and sides of each rabbit, squares of skin measuring 3 × 3 to 4 or 5 cm were carefully cleared of hair so that four to five rows with five to seven squares were obtained, and the skin looked like a chessboard (Fig. 2). The skin tests were made 24 h after depilation. The extracts from grains or fungi were applied by a platinum loop or by micropipet to the skin of the rabbit twice within a 24-h interval. Two squares of each rabbit served as a control and were treated with ethanol extract of autoclaved normal wheat grain not infected by any fungus. The reaction was recorded for 48 h, but the rabbits were kept under observation for at least 6 to 8 days after the first application.

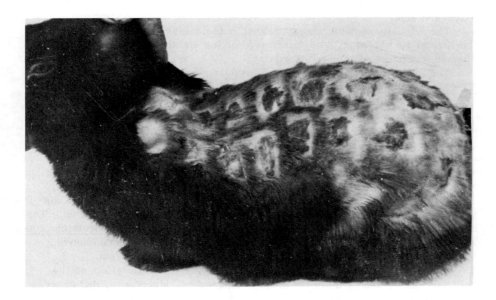

Fig. 2.   General view of rabbits treated with toxic extracts of <u>F</u>. <u>poae</u>, <u>F</u>. <u>sporotrichioides</u>, and <u>F</u>. <u>sporotrichioides</u> var. <u>tricinctum</u> 48 h after application. The white squares on the left are controls.

Prior to being subjected to the experimental application, each rabbit was given
two control treatments, one with an extract of known toxicity and another with
the extract of unaffected grain [68, 69, 73]. During the experiment stiff
collars of cardboard prevented the rabbits from licking the test squares.

3.10.3.2   The Reaction of Skin Test

There were various types of skin reactions, the evaluation of which was based
on their principal components: leukocytic film, edema, hemorrhage, and
necrosis.
    The leukocytic reaction was characterized by the formation on the skin
surface of a whitish cream, an easily detachable film which consists of infiltration of mass leukocytes accumulating in the horny layer of the epithelium
or dermis. The intensity of the leukocytic components was estimated according to the massiveness of the superficial film.
    The edema or acute edema reaction sometimes caused severe inflam-
matory changes in the epidermis; the intensity of the edema was assessed by
the thickness of the skin fold. In this case there was no leukocytic film, or
sometimes the leukocytic infiltration of the dermis was accompanied by edema
and necrosis of the skin.
    The appearance and intensity of hemorrhage reaction was connected also
with inflammatory changes in the skin and estimated by the quantity of visible
extravasations.
    The intensity of necrosis, as judged by the massiveness of the scab and
the time of its shedding, was determined on the eighth day; the other components were recorded on the third day.
    The intensity of the skin reaction produced by each culture of F. poae and
F. sporotrichioides and others was assessed on a scale of grades by Joffe
[68, 69] and Joffe and Palti [79].
    The presence of edema, hemorrhage, and necrosis was regarded as
evidence of marked toxicity of the fungus or of the grain samples. A pronounced leukocytic film, unaccompanied by any of the other components, was
considered as an indication of the toxicity of the experimental material; a
reaction consisting of a thin leukocytic film, scattered vesicles, reddening,
and desquamation was assessed as weak and doubtful [68, 69, 72, 73, 79].
    Skin reactions caused by the action of Fusarium extracts were thus dis-
tinguished both by their external appearance and by histological changes. The
areas of skin which were in contact with toxins of F. poae, F. sporotrichioides,
and F. sporotrichioides var. tricinctum were fixed with 4% formaldehyde.
Paraffin sections were cut and stained with hematoxylin and eosin.
    The toxin of Fusarium extracts often had a general toxic effect on the
rabbits, manifested by a loss of appetite and weight, sleepiness, and changes in
the blood composition and in the organs; in not infrequent cases, the animals
died after the administration of the toxin.

3.10.3.3   Other Tests

Davydova [29] and Mironov and Davydova [123] found that the skin of cats, sheep, dogs, cows, and horses was also sensitive to Fusarium toxins and overwintered cereals. Similar results were obtained by pipetting toxic extracts on the conjunctiva of rabbits. The skin of pigs, guinea pigs, white mice, and rats was less sensitive; the skin of man was altogether insensitive [124, 125]. In contradistinction to these results, Schoental and Joffe [172] established that small doses applied to the skin of mice and rats have a strong local cytotoxic effect.

Other tests were also suggested for the recognition of toxic grain. Thus, Kretovich and Skripkina [93] and Misuhstin et al. [129] proposed a quick test which was based on the observation that the ethanol extract of toxic grain of prosomillet, wheat, and other crops suppressed the fermentative power of Saccharomyces cerevisiae [91]. Elpidina [34] suggested a phytotoxicological test in which a drop of toxic grain or ethanol extract of Fusarium poae on leaves caused necrotic dark brown spots on plant tissue of leaves and the wilting of the plant. Joffe [74] also put different plants (tomatoes, beans, peas) into liquid media containing toxic metabolites of F. poae and F. sporo-trichioides. After 4 h the turgor had already decreased and the plant had begun to droop. After 20 to 24 h the plants were completely dry. Seedling mortality was caused also by inoculation of fungi of the Sporotrichiella section onto field and vegetable crops [74]. Toxic extracts of these Fusarium species caused more severe changes than toxic extracts of prosomillet and wheat over-wintered in the field.

Elpidina [34, 36] detected that 2 drops of the toxic extract of prosomillet infected by Fusarium poae reduced the antitoxic power of diphtheria antiserum, and the toxin named poin obtained from this Fusarium inhibited the growth of Ehrlich's adenocarcinoma and Croker's sarcoma [37-40].

Tkachev [195] suggested that dough made of flour from overwintered grains, when boiled in milk, coagulated the milk; this did not happen with meal of normal nontoxic overwintered grain. Manoilova [110], Kozin and Yershova [88], and Olifson [141, 142] proposed color change methods in the extract of toxic overwintered grain and toxic Fusarium strains of the Sporo-trichiella section.

3.10.3.4   Bioassay with Toxic Fusarium

In feeding experiments of various animals, we used different bioassay methods. The culture filtrates, mycelium (dry fungi), infected grains, ethanol, or ethered extract of toxic F. poae, F. sporotrichioides, or F. sporotrichioides var. tricinctum were administered per os and culture filtrate and crude extract were administered subcutaneously or intraperitoneally in various doses to laboratory animals [68, 69, 73, 75]. We also used gastric fistulas in experiments with dogs [61] and intragastric intubation with mice and rats [172].

The following animals were used in our studies for the biological assay
test with toxic strains of F. poae, F. sporotrichioides, and F. sporotrichioides
var. tricinctum (including observations on loss of weight, mortality, and
histopathological changes in the organs and tissues): frogs, mice, rats, guinea
pigs, rabbits, cats, chickens, ducklings [8, 68, 69, 75, 77, 172], and also
some protozoa [31, 32, 68]. All the animals that died were autopsied and
tissues were stained with hematoxylin and eosin for histological investigation.

Sarkisov [167, 169] carried out experiments by the per os method on
mice, rats, guinea pigs, rabbits, dogs, and cats as well as on horses, cattle,
sheep, and pigs. He fed them toxic overwintered grain infected by F. sporotrichioides. Bilai [17-19, 21] and Pidoplichka and Bilai [149] fed young
rabbits and guinea pigs grain infected by F. poae and F. sporotrichioides and
used aqueous extracts by subcutaneous injections for the assessment. Rubinstein [156-161] and Rubinstein and Lyass [162] carried out studies on feeding
mice, rats, and monkeys per os with grain infected by F. sporotrichioides.
Getsova [49] used intraperitoneal injections on mice with toxins of ATA.
The effect of Fusarium toxins in animals has been reviewed by Joffe [68, 69,
73, 75].

3.10.4   Fungi and Toxins

Many mycologists studied the mycoflora of the toxic overwintered cereal crops
in relation to ATA and have isolated different kinds of fungi from the toxic
grains of wheat, barley, rye, oats, and especially prosomillet. Some of these
fungi caused plant diseases [74, 75, 77] and others produced toxic metabolites
in animals and man [68-75, 79].

3.10.4.1   Fungi

Murashkinskij [130] isolated some strains of Alternaria (see Sec. 1.1.2 for
key to genera) and Fusarium from Siberian toxic wheat grain, and Sirotinina
[181] isolated some Alternaria from toxic prosomillet of the Saratov district.
Kvashnina [95] isolated 1227 strains belonging to 83 species, among which
F. sporotrichioides, Aspergillus caliptratus (Sec. 1.2.3), Phoma spp. (Sec.
1.8), Hymenopsis spp. (Sec. 1.8), were very toxic [167, 169], from 107
samples (prosomillet, wheat, rye, oat, barley, pea, and sunflower) gathered
in the Altai territory, Bashkir and Tatar A.S.S.R., Byelorussian S.S.R., and
the Ivanov, Saratov, Kuybyshov, Tambov, Chkalov, and Yaroslav districts.

Pidoplichka and Bilai [149] and Bilai [19] examined about 1400 strains
belonging to 23 genera and 160 species isolated from 765 grain samples (of
prosomillet, wheat, oat, buckwheat) obtained in Bashkir A.S.S.R. and
Ukrainian S.S.R. The most toxic isolates were Mucor hiemalis and M. albo-
ater together with Piptocephalis freseniana, Mortierella polycephala, M.
candelabrum var. minor, F. poae, F. sporotrichioides, F. lateritium,
Gliocladium ammoniophilum, and Trichoderma lignorum (see Sec. 1.1.2 for
key to genera). The present author isolated 3849 fungal cultures belonging to

42 genera and 192 species [68, 69] from more than 1000 samples of overwintered and normal grains (prosomillet, wheat, barley, oats, rye, buckwheat,
and sunflower) in the Orenburg district.

The number of genera found on samples of overwintered plants was propor-
tionately much larger than the number on summer-harvested samples. Thus,
on overwintered cereals, the genera represented by numerous isolates were
Penicillium, Fusarium, Cladosporium, Alternaria, and Mucor. On normal
samples only the three genera Penicillium, Mucor, and Alternaria were present
as a rule, whereas Fusarium and Cladosporium were found in only a few cases,
or were absent altogether.

In summary, one can conclude that the wintering of cereals under snow
resulted in the formation of a rich and heterogeneous mycoflora, and in an
active increase of fungi of the genera Penicillium, Alternaria, Mucor,
Fusarium, and Cladosporium. The degree of infection was much higher in
overwintered cereals than in those harvested normally.

For the evaluation of the role of Fusarium in toxin production it seemed
necessary to study the toxicity of isolated cultures. Among 501 Fusarium
cultures isolated from the substrata just mentioned, 35.7% showed toxicity of
varying degrees, 22.4% were highly toxic, and 13.3% slightly toxic (Table 1).
Highly and mildly toxic cultures were much less common in Cladosporium, and
still less in Mucor, Alternaria, and Penicillium. No toxic cultures were found
from grain or from vegetative parts of summer-harvested plants.

Further investigations were aimed at assessing the genera most likely to
produce toxins. It was assumed that the toxicogenic properties of different
genera of fungi might be estimated from the frequency of their occurrence on
overwintered cereals and from the appearance among them of highly toxic

Table 1   Genera of Fungi Associated with Toxin Production in Overwintered
           Grain

| Genus | Isolates[a] | | | Species | | |
	Total no.	Highly toxic (%)	Mildly toxic (%)	No. isolated	Highly toxic (%)	Mildly toxic (%)
Fusarium	501	22.4	13.3	25	60	28
Cladosporium	480	5.4	8.5	15	60	20
Alternaria	506	2.8	5.3	6	30	0
Penicillium	830	1.6	3.8	36	22	33
Mucor	335	3.0	7.2	18	33	22

[a]Highly toxic isolates were also found in Piptocephalis freseniana (with Mucor
albo-ater), Trichoderma lignorum, Rhizopus nigricans, Trichothecium
roseum, Thamnidium elegans, Verticillium lateritium, and Actinomyces
griseus.

Table 2   Toxic Fungi Isolated from Overwintered Cereals, Soils, and Summer-Harvested Cereals[a]

Fungi	Overwintered cereals			Soils			Summer-harvested cereals
	Toxic	Mildly toxic	Non-toxic	Toxic	Mildly toxic	Non-toxic	Nontoxic
Phycomycetes							
Mucor corticola Hag.	1	1	19		1	6	8
M. hiemalis Wehm.	3	4	34	2		13	20
M. humicola Raillo	1	1	9		1	8	
M. racemosus Fres.	2	3	33		1	20	46
Piptocephalis freseniana D. B. et W. with Mucor albo-ater Naum.	1	3	4			14	
Rhizopus nigricans Ehr.	1	2	2		1	16	54
Thamnidium elegans Link	1	3	33			11	6
Ascomycetes							
Penicillium brevi-compactum Dier.	2	1	23				
P. chrysogenum Thom	1	1	40			8	7
P. cyclopium Westl.	1	1	24			3	
P. nigricans Bain.	2	2	34	1	1	11	
P. notatum Westl.	1	2	23			7	19
P. steckii Zal.	2	1	11			7	
P. umbonatum Sopp	1	1	25			4	
P. viridicatum Westl.	1	1	27		1	4	
Fungi imperfecti							
Alternaria humicola Oud.	2	4	40			9	3
A. tenuis Nees.	10	21	350	2	2	58	81
Cladosporium epiphyllum (Pers.) Mart.	8	11	73	2	2	25	2
C. exoasci Link	2		15			3	1
C. fagi Oud.	4	1	33			9	
C. fuligineum Bon.	1	1	17			6	
C. gracile Cda.	3		19			4	
C. herbarum (Pers.) Link	2	12	35		1	24	11
C. molle Cke.	2		29			2	
C. penicillioides Preuss	1	4	36		1	5	2
C. pisi Cug. et March.		4	14			8	
Trichoderma lignorum Tode (Harz)	1		18			10	2

Table 2 (Continued)

Fungi	Overwintered cereals			Soils			Summer-harvested cereals
	Toxic	Mildly toxic	Non-toxic	Toxic	Mildly toxic	Non-toxic	Nontoxic
Trichothecium roseum Link	2	3	26			5	
Verticillium lateritium Rabenh.	1	1	8				4

[a]Remarks: Occasional toxic isolates were also found in Phycomycetes: Chaetomium brefeldi, Mucor dispersus, M. fumosus, M. globosus, M. griseo-ochraceus, M. heterosporum, M. murorum, M. oblongisporus, M. silvaticus. Ascomycetes: Aspergillus fumigatus, A. niger, Penicillium auratio-virens, P. biourgeianum, P. citreo-roseum, P. crustosum, P. cyaneo-fulvum, P. griseo-roseum, P. howardii, P. martensi, P. micrznskii, P. palitans, P. purpurogenum, P. westlingi. Fungi imperfecti: Botrytis cynerea, Clado-sporium graminum, C. grumosum, C. spherospermum, Coremium glaucum, Gliocladium penicillioides, Gonotobotrys flava, Macrosporium commune. Actinomycetes: Actinomyces globisporus, A. griseus.

strains. From toxic and highly toxic cultures, 13 genera were isolated, and from mildly toxic, 17 genera. Most toxic fungi belonged to the genera Alternaria, Mucor, Penicillium, and especially Fusarium and Cladosporium, each of these being represented by many species [68, 69]. Fusarium poae and F. sporotrichioides, both of very common occurrence on overwintered cereals in all those parts of the Orenburg district from which material had been collected, were represented in the majority of cultures.

Tables 2 and 3 list the different species, especially the Fusarium fungi, which had been isolated from overwintered and summer-harvested cereals and soil. The results (Table 3) showed that the group most frequently associated with the overwintered cereal grains and which caused ATA was Fusarium of the Sporotrichiella section, principally F. poae and F. sporotrichioides. According to Joffe [68, 69, 72-75], Joffe and Palti [79], Sarkisov [167, 169], Bilai [19], Pidoplichka and Bilai [149], Rubinstein [155, 157], and Rubinstein and Lyass [162], one of the characteristic biological properties of F. poae and F. sporotrichioides compounds was their inflammatory and irritative action on rabbit skin.

Overwintered cereal grains that were toxic included 61 F. poae (Peck.) Wr., 57 strains of F. sporotrichioides Sherb., and F. sporotrichioides var. tricinctum toxic 3, nontoxic 25 [68, 69] (Table 3). Sarkisov [167] found 14

Table 3   Toxicity of _Fusarium_ Fungi Isolated from Overwintered Cereals, Their Soil, and Summer-Harvested Cereals

Fungi	Overwintered cereals			Soil			Summer-harvested cereals
	Toxic	Mildly toxic	Non-toxic	Toxic	Mildly toxic	Non-toxic	Nontoxic
Section _Arachnites_							
Fusarium nivale (Fr.) Ces.		3	20				
Sect. _Sporotrichiella_							
F. _poae_ (Pk.) Wr.	44	17	2	2	2		
F. _sporotrichioides_ Sherb.	42	15	4	2	2		2
F. _sporotrichioides_ Sherb. var _tricinctum_ (Cda.) Raillo	2	1	19			5	
Sect. _Roseum_							
F. _avenaceum_ (Fr.) Sacc.	3	3	26			10	3
F. _arthrosporioides_ Sherb.		1	7			2	5
Sect. _Arthrosporiella_							
F. _semitectum_ Berk. et Rav.	2	3	26				
Sect. _Lateritium_							
F. _lateritium_ Nees	2	2	24		1	3	
Sect. _Liseola_							
F. _moniliforme_ Sheld.	1	3	22		1	10	
Sect. _Gibbosum_							
F. _equiseti_ (Cda.) Sacc.	4	1	19			6	18
F. _equiseti_ (Cda.) Sacc. var. _acuminatum_ (El. et. Ev.) Bil.	1		5				
F. _equiseti_ (Cda.) Sacc. var. _caudatum_ (Wr.) Joffe	2	2	17		1	3	9
Sect. _Discolor_							
F. _graminearum_ Schw.		1	2				
F. _sambucinum_ Fuckel.	1	1	14				
F. _culmorum_ (W. G. Sm.) Sacc.	2	1	13				
Sect. _Elegans_							
F. _oxysporum_ Schl.	1	2	16		1	13	2
Sect. _Martiella_							
F. _solani_ (Mart.) Sacc.		3	16			5	
F. _javanicum_ Koord.			8				5

Fig. 3.   Edematous hemorrhagic necrotic reactions on rabbit skin 24 h after application of P. poae (left), F. sporotrichioides (middle), and control (spot on the right).

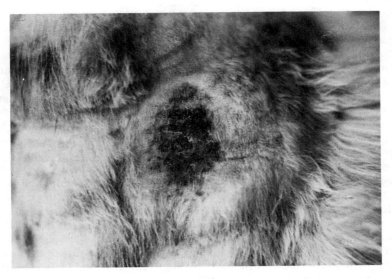

Fig. 4.   Edematous necrotic reaction on rabbit skin 72 h after application of F. sporotrichioides var. tricinctum (right) and control (left).

F. poae and 32 F. sporotrichioides, and Bilai [19] 34 F. poae and 36 F. sporotrichioides. Isolates of F. poae, F. sporotrichioides, and F. sporotrichioides
var. tricinctum generally caused an edematous hemorrhagic reaction followed
by severe necrosis (Figs. 3 and 4).

   A relationship was found between the nature of the toxic Fusarium cultures
and the toxicity of the samples from which they had been isolated. Some of the
Fusarium cultures caused reactions on the skin of rabbits which were
analogous to those produced by the action of toxic cereals which had passed
the winter under snow [73].

   A detailed description of morphological and cultural properties of the toxic
fungi of the Sporotrichiella section has been given by Joffe [68, 69, 73, 75].
(See Section 1.4.5.)

3.10.4.2   Toxins

The literature published in the U.S.S.R. on the chemistry of overwintered
grain toxins and on the structure and chemical properties of F. poae and F.
sporotrichioides [15, 59, 83, 89, 90, 92, 93, 121, 132, 133, 136, 138, 140,
142-146, 190, 209], has recently been reviewed by Joffe [73, 75].

   At the University of Wisconsin, recent studies were conducted which were
associated with outbreaks of toxicosis in animals ingesting moldy corn. The
results of these studies were in contradistinction to those of Soviet researchers.
Many species of fungi were isolated from toxic corn samples, among them
F. tricinctum, which was one of the more toxic strains.

   Several highly toxic metabolites, spiroepoxy derivatives of the trichothecene group, were isolated from F. tricinctum. In addition, many studies were
conducted in the United States and in Japan which cast doubt on studies carried
out in Russia [142], namely the role of steroid materials being responsible for
the toxic action of F. poae and F. sporotrichioides.

   The following interesting works were carried out in the United States and
Japan with F. tricinctum: Goldfredsen et al. [51]; Gilgan et al. [50]; Bamburg
et al. [10-12]; Bamburg and Strong [13, 14]; Marasas et al. [112]; Smalley
et al. [183]; Burmeister [22]; Burmeister and Hesseltine [23]; Burmeister
et al. [24]; Mirocha and Pathre [116]; Grove et al. [57]; Yates et al. [203,
204]; Hsu et al. [62]; Scott and Somers [173]; Tookey et al. [197]; and Ueno
et al. [198, 199]. These studies were carried out with toxins of fungi obtained
from the U.S.S.R. and with U.S. strains determined as F. tricinctum (Corda)
Sacc. emend. by Snyder and Hansen. We have reason to assume, according
to mycological standards, that they should more properly be called F. poae
(Peck.) Wr. and F. sporotrichioides Sherb. In the U.S. works the toxins were
found to be diacetoxyscirpenol and T-2 toxin (Sec. 2.5.1.3), which are ses-
quiter penoid and a derivative of the trichothecene group.

   Bamburg and Strong [14] and Mirocha and Pathre [116] have concluded
that ATA is more likely caused by the trichothecene toxins rather than by the
steroidal toxins reported by Olifson [138-140, 142, 143] and Olifson et al.
[146]. Also, in Japan Ueno et al. [198, 199] showed by chemical analysis
that in F. poae and F. sporotrichioides there were present trichothecene-type

toxins (T-2 toxin, neosolaniol, and a trace of HT-toxin; Sec. 2.5.1.3). A large amount of butanolide (Sec. 2.5.1.3) was also isolated from the metabolites of F. sporotrichioides [199].

In view of the fact that the problem of ATA was a matter of great interest, and that neither the U.S. nor the Japanese investigators who had worked with F. tricinctum had been able to detect the toxins isolated by Russian scientists, we suggest that the uniform classification of toxic strains is very important and essential for an advanced discussion on the mycotoxin problem.

This study will be carried out with authentic strains of F. poae and F. sporotrichioides isolated from overwintered grain which were responsible for alimentary toxic aleukia. However, the subject is now being reinvestigated at the School of Pharmacy of the Hebrew University in cooperation with Dr. B. Yagen. A detailed discussion cannot be undertaken, pending the results of this study.

3.10.4.3   Taxonomic Problems of the
            Fusarium of the Sporo-
            trichiella Section

The taxonomy of species of the Sporotrichiella section may once have appeared a subject of purely academic interest, but this no longer holds true. Ever since species of this section have been proven to induce serious disorders in humans, who died after consuming overwintered grain [68, 70-73, 75], and various diseases in animals [23, 24, 57, 62, 111, 197, 198, 203, 204] the true identity of these species has become of acute importance. Establishment of this identity will enable us to relate, at least from a taxonomic angle, a large body of toxicological work carried out in recent years, mostly in the United States and Japan, to the earlier work carried out principally in the U.S.S.R.

There is a problem in the taxonomy which derives from the following situation: The great majority of taxonomists, since Wollenweber and Reinking [202], have recognized several distinct species in the Sporotrichiella section [48, 52-54, 76, 151]. Moreover, Joffe [68, 70, 72-75] and Seemüller [174] have proved that these species also differ in their toxicological properties from what has generally been called F. tricinctum, producing lower numbers of isolates with toxic properties, and F. poae and F. sporotrichioides, producing such isolates in large numbers.

However, in the United States, Snyder and Hansen [185], for reasons never adequately explained, decided to group all the species in the section together under F. tricinctum. Since most of the toxicological work in recent years has been carried out by researchers not particularly interested in the taxonomic identity of the isolates with which they worked, Snyder and Hansen's nomenclature was adopted uncritically. We do not accept Snyder and Hansen's grouping of all the species together under the name of F. tricinctum. Our concepts approximate those of Gordon [52-54] and Seemüller [174] with some slight variations, and we distinguish between two species of F. poae (PK) Wr. and F. sporotrichioides Sherb., and two varieties, F. sporotrichioides var. tricinctum and F. sporotrichioides chlamydosporum (Fig. 5).

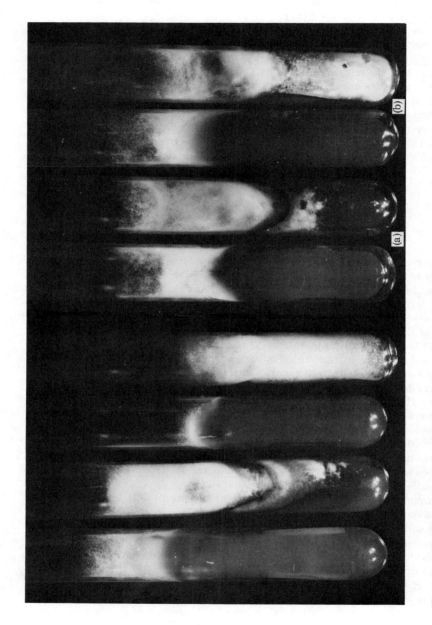

Fig. 5   Fusarium sporotrichioides (left) and F. poae (right); F. sporotrichioides var. tricinctum (a) and F. sporotrichioides var. chlamydosporum (b) on medium of potato-dextrose-Agar (PDA).

Table 4 Cultural, Morphological, and Toxicological Characteristics of the Species and Varieties in the Section Sporotrichiella

	F. poae	F. sporotrichioides	F. sporotrichioides var. tricinctum	chlamydosporum
Aerial mycelium				
Color	White, yellow, red or red-brown	White, whitish, rose, or red	White, carminered to purple	White, light yellow to carmine-brown, or intensive rose
Consistency	Felted, somewhat powdery	Downy	Weely	Downy, flocculose
Growth rate	7–8 cm	4–4.5 cm	2.7–3 cm	3–3.5 cm
Microconidia				
Shape	Oval, globose spherical with basal papilla, rarely pear-shaped	Globose, pear-shaped, ellipsoid, elongate	Lemon- and also pear-shaped, oval	Spindle-shaped or elongate, rarely oval-ellipsoidal
Occurrence in aerial mycelium	Single or in false heads	Singly, rarely in chain or false heads	In false heads	Singly, rarely in groups
Measurements (μm)				
0-septate	6.8–9.5 × 4.6–8.2	6–11 × 4.0–7.0	7–9.5 × 2.8–7.2	7.5–10.5 × 2.6–3.2
1-septate	10.5–15 × 4–7.4	9.20 × 3,8–7.5	10–19 × 3–5.5	11–14 × 3.0–3.8
Abundance in relation to macroconidia	More abundant than macroconidia	Often as numerous as macroconidia	More abundant than macroconidia	Usually more numerous than macroconidia
Macroconidia				
Shape	Falcate, sometimes lightly curved without foot cell	Falcate to curved with or without foot cell	Falcate, elliptical or more strongly curved, with well-marked foot cell	Curved with narrowly painted apex, well-marked foot cell

Table 4   (Continued)

	F. poae	_F. sporotrichioides_	_F. sporotrichioides_ var. tricinctum	chlamydosporum
Measurements (μm)				
3-septate	19–38 × 3.5–6.0	28–35 × 3.2–4.6	23–45 × 3.4–4	28–35 × 3.2–3.8
5-septate	19–38 × 3.5–6.0	37–44 × 3.5–4.8	34–51 × 3.6–4.4	38–44 × 3.5–4.5
Frequency of				
3-septate	Preponderant	Numerous	Preponderant	Preponderant
5-septate		Numerous	Rare	Rare
Location	Aerial mycelium only	Aerial mycelium sporodochia, rarely in pionnotes	Aerial mycelium, sporodochia	Aerial mycelium rarely in sporol sporodochia
Chlamydospores	Intercalary, singly or in pairs or chains; rare, sometimes absent; hyaline to light brown	Intercalary, singly or in pairs, knots, or chains; sometimes terminal; smooth-walled, hyaline to light brown	Intercalary, singly or in chains; rarely terminal; smooth-walled, brown	Terminal, singly or in pairs or chains; smooth-walled or rough, brown
Plectenchymatous sclerotia	Absent	Occasionally present	Present	Absent, or occasionally present in young cultures
Stroma on PDA	Red or ochre-yellow sometimes violet, rarely colorless	Blood-red, yellow, purple, brown, or light carmine	Carmine-purple, ochre-brown, rarely colorless	Red, brown-red, carmine, purple, or light violet
Toxicity to rabbit skin				
Frequency of toxic isolates	High	High	Low	Low to very low

Criteria for distinguishing between species of the <u>Sporotrichiella</u> section
have recently been discussed by Joffe [76] and are summarized in Table 4.

3.10.5   Etiology

The seasonal occurrence of ATA, its endemicity, and the composition of the
affected population suggested the importance of climatic and ecological factors
in the outbreaks of the disease.

Surveys in the affected areas showed that the major nutrients of the patients
consisted of the cereal grains wheat, prosomillet, barley, rye, oats, and buckwheat. It was thus assumed that there was a correlation between the food of
these rural peasants and development of the disease.

The peak outbreaks occurred in the spring when overwintered grains which
had been under snow cover for the entire autumn and winter were consumed by
the rural peasants. This led investigators to look for changes in the grains as
a result of their being under snow throughout the winter. Considerable changes
were found in the mycoflora of the overwintered cereals, which suggested that
ATA was caused by certain fungi that developed well under cover of the snow.

3.10.5.1   General Picture of
            Mycological Study

Initially it was thought that prosomillet was the most dangerous source of ATA,
since many people fell ill after eating this grain. This cereal crop was an
extensive source of the disease; it was widely grown in the Orenburg district
and in other parts of the Soviet Union, and ripened late, producing a very great
harvest so that large amounts were left to overwinter in the field. Later it was
shown that wheat and barley were the main causes of the disease, as well as
oats, rye, and buckwheat [71].

All these cereals were left unharvested during the winter under snow, and
the climatic, meteorologic, and ecologic conditions were favorable for
enhancing toxic properties. According to Joffe [65, 71, 73] the great number
of deaths in the Orenburg district and in other places in the U.S.S.R. were
closely associated with the consumption of toxic overwintered grain infected
with <u>Fusarium</u> fungi. Researchers who worked with the ATA problem suggested
that this disease was associated with a really toxic origin, especially with
overwintered grains contaminated by toxic fungi. This hypothesis was indeed
correct and important in the etiology of ATA and was supported by many investigators [19, 33, 64, 68-75, 95, 127, 149, 155, 156, 162, 167-169]. The
studies on the overwintered and normal grains, mycoflora, and the toxicity of
the various isolated species enabled us to understand more clearly the etiology
of ATA [64, 65, 68, 69]. The fungi most frequently associated with the toxic
grain causing ATA belonged to the <u>Sporotrichiella</u> section of the genus <u>Fusarium</u>
and included the following species: <u>F</u>. <u>poae</u>, <u>F</u>. <u>sporotrichioides</u>, <u>F</u>. <u>sporo-
trichioides</u> var. <u>tricinctum</u>, and <u>F</u>. <u>sporotrichioides</u> var. <u>chlamydosporum</u>
[76, 80].

The causes of ATA were identified as toxic fungi [68, 69]. The identification followed these steps: The regional offices of the U.S.S.R. Ministry of Health gave notification of death caused by ATA. Samples of the grain and products consumed by the deceased were taken from their homes. These samples were carefully investigated with the help of mycological analysis (Figs. 6 and 7) and the fungal extracts obtained were then examined by the skin test method on rabbits, and at the same time were fed to animals. The toxins produced by F. poae and F. sporotrichioides, which cause ATA, were investigated in detail on various animals by Joffe [64, 68, 69, 72, 73, 75]. When these animals exhibited symptoms resembling those observed in man, the toxic principle was considered to be inherent in the samples.

The toxins of F. poae and F. sporotrichioides were also isolated from different overwintered grains and soil collected from various lethal fields of the Orenburg district.

In addition to all these investigations, we carried out experiments from 1943 to 1949 on 39 variously treated trial plots in the Orenburg district. These experiments contributed much to the determination of the conditions for toxin production in overwintered grain [71].

3.10.5.1.1   Materials and Methods. All the isolated Fusarium species were initially cultured in Petri dishes on potato-dextrose-agar (PDA) at 18°C, then in test tubes, and finally on various liquid and solid media at different temperatures (1, 6, 12, 18, 24, 35, and 40°C) for a period of 6 to 76 days, and sometimes from +4 to -5, -7, and -10°C on natural media for a period of 30 to 90 days or more. When using natural media, prosomillet, wheat, or barley grains were prepared in 150-, 250-, 500-ml, or 1-liter flat flasks (Fig. 8) containing 10, 30, 50, or 100 g grain, and 15, 45, 75, or 140 ml tap water, respectively. The agar cultures were placed in a 3-liter flask on whose walls a layer of agar had been deposited by rolling. A variety of liquid media were used [66-69, 74, 75, 79]. The cultures were incubated in 1000 or 2000 ml Fernbach flasks containing 400 to 800 ml of liquid medium, respectively. The solid and liquid media were autoclaved at 0.5 to 1 atm for 20 to 30 min or were sometimes steam sterilized two or three times for 1 h. The natural, agar, and liquid media were inoculated with fungal spores. After incubation they were taken out and dried overnight at 45 to 50°C. A portion of the powdered mycelia and culture filtrate was then extracted with 70 to 96% ethanol or ether and concentrated in a vacuum rotary evaporator. After cultivation the liquid media were filtrated with filter paper or passed through a Seitz filter. The pH of the media varied depending on the time of cultivation, and the kind of medium.

3.10.5.1.2   Conditions for Toxin Formation. Conditions for toxin production were studied in the laboratory and in the field [71]. Thirty-nine experimental plots of land were set up under normal field conditions in different counties of the Orenburg district and examined for 6 years. The plots varied in size from 100 m^2 to 0.5 ha. The experimental crops included the most commonly grown cereals, namely millet, wheat, and barley.

Fig. 6   Sample of overwintered prosomillet affected by F. sporo-trichioides consumed by two members of the family who died.

Fig. 7   Wheat ear overwintered in the field which caused death of three persons after consumption.

Fig. 8   F. poae I and F. sporotrichioides II grown on wheat in 1-liter flat flasks.

At harvest time one-half of the crop of each experimental plot was cut and arranged in stacks, while the other half was left uncut. Samples of the stacked cereals were taken from the upper layer of the stack, as well as from the bottom layer, at soil level. Plant and soil samples were usually collected from the experimental plots twice a month, from August or September of each year to the following May. The cereal samples were threshed and then dried at 45°C; the soil samples were similarly dried. Samples of dried and threshed grain, of vegetative parts (stems, leaves, ears, panicles, husks, etc.), and of soil were soaked in ether or alcohol. After 3 to 5 days soaking with repeated shaking, the ether or alcohol was driven off in a distilling apparatus. The residue obtained after evaporation was assayed for toxicity by application to the shaved skin of a rabbit. This method of testing has been fully described in the section on the mycoflora of overwintered cereals and their toxicity.

At the beginning of each month, a summary was made of the preceding month's weather conditions for each of the experimental plots. These records included detailed temperature, moisture reading, the amount of precipitation, depth of freezing of the soil, and average depth of the snow cover [71].

The material from this 6-year investigation of the experimental fields consisted of 1535 cereal and soil samples which were assayed for toxicity.

Concurrent investigations were conducted during the years 1943 to 1949 on more than 1500 samples of overwintered cereals collected from all counties of the Orenburg district, on control samples of cereals harvested in the normal summer harvest period, and on soil samples. The material for investigation consisted of millet, wheat, barley, oats, and buckwheat as well as occasional samples of sunflower seeds, acorns, and legumes gathered in the fields.

Both the collated material for the period 1943 to 1949 and the results obtained from the same plot in different years indicated that the degree of toxicity of overwintered cereals varied according to the year. On the basis of the data, it can be assumed that only in certain years were conditions favorable for the accumulation of toxic substances in cereals overwintering under the snow.

The years 1943 and 1944 were characterized by exceptionally potent toxicity of the grains. Samples from this period almost invariably produced a positive skin reaction. Particularly toxic samples were collected from different counties in the spring of 1944. Of the tested samples, 75.6% exhibited a strong toxic activity whereas in the remaining 24.4% toxicity was less pronounced. The number of toxic samples from the spring 1944 experimental plots exceeded the number of toxic samples collected during the entire period between the autumn of 1944 and the spring of 1949 [71, 73].

Samples collected in 1945, 1948, and 1949 were mostly nontoxic. The years 1946 and 1947 occupied an intermediate position, in that the toxicity of samples investigated in those years, even though not high, was quite significant [71].

3.10.5.1.2.1   Relation between the Incidence of ATA and the Toxicity of Cereals. The disease appeared in 1942, and increased considerably in 1943 and especially in 1944, the peak year, when the population in the Orenburg district of Russia suffered enormous casualties.

In 1943 the percentage of toxic samples was high but was not paralleled by a high incidence of disease: This was because food in Russia was not yet so scarce as to force large numbers of people to collect overwintered grains. However, in 1944 the high prevalence of toxicity in the samples coincided with extreme scarcity of food, and large parts of the population could subsist only by searching for grain overwintered in the fields; this resulted in a very high incidence of disease in that year. In the following years, the toxicity of the samples decreased and the food situation improved, so that the incidence of the disease was greatly reduced.

3.10.5.1.2.2   Persistence of Toxicity in Stored Grain. With a view toward determining whether the toxicity of overwintered cereals is modified by prolonged storage, strongly toxic samples of millet, retained from the years 1943 to 1944, were reinvestigated in 1949. The results demonstrated that the toxic ingredient contained in the grain was not affected by 6 or 7 years of storage and that toxicity persisted unimpaired [70, 73, 75].

However, concurrent mycological investigations of these samples failed to isolate F. poae and F. sporotrichioides, which had obviously perished. These findings may account for the difficulties encountered in attempts to isolate toxic fungi from cereal grains after a prolonged period of storage.

3.10.5.1.2.3   Sources of Toxins Causing ATA.   Large-scale investigations were carried out from 1943 to 1949 on samples from various plants and plant organs and from soil, all collected from experimental cereal fields in the district in which ATA outbreaks in the U.S.S.R. were heaviest [71, 73, 75]. The results of these investigations are presented separately for 1943/1944, when the outbreak was disastrous, and for the remaining years when there were lighter outbreaks (Table 5).

The data show the enormous rise in the percentage of toxic samples in the year in which outbreaks of the disease were most severe. In the subsequent years, of lesser disease severity, wheat samples were more frequently affected than those of millet or barley. Grains were always more severely toxic than vegetative parts; in soil the percentage of slightly toxic samples was particularly high.

3.10.5.1.2.4   Seasonal Effects.   The observations showed that, in general, toxin formation in overwintered cereals took place during the autumn-winter-spring period. It remained to be seen during which of these seasons the production of toxins was largest. The largest number of toxic samples was detected in spring. In winter, toxic samples were less frequent, whereas the number of toxic samples recorded in autumn amounted to less than half that observed in the spring [68, 73].

Table 5   Toxicity of Samples of Various Cereals, Plant Organs, and Soil in the Year of Heaviest ATA Outbreaks (1943/1944) and in Five Subsequent Years with Lower Incidence of Disease

Source of samples	No. of samples examined	Percentage of samples		
		Toxic	Slightly toxic	Nontoxic
1. Species of cereal				
1943/1944 Millet	243	60.8	15.2	27.9
1944/1949 Millet	420	2.6	5.2	92.2
Wheat	415	4.6	4.6	90.8
Barley	255	1.9	5.5	92.6
2. Plant parts and soil				
1943/1944 Vegetative parts	134	39.6	20.1	40.3
Grains	109	78.0	9.2	12.8
1944/1949 Vegetative parts	510	1.9	4.3	93.8
Grains	539	4.6	8.1	89.2
Soil	285	2.8	8.4	88.8

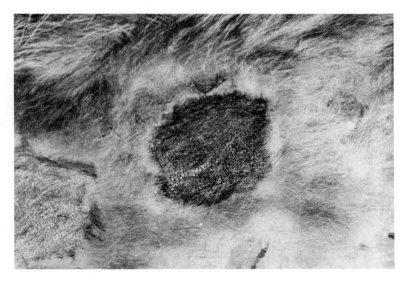

Fig. 9   Necrosis and moderate edema of the rabbit skin 72 h after application
of toxic overwintered prosomillet extracts.

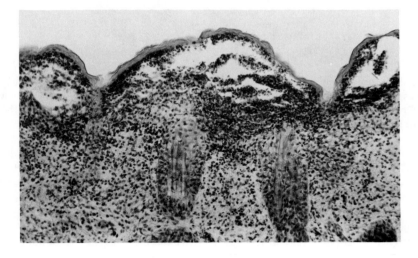

Fig. 10   Extract of overwintered prosomillet (from experimental field).
Section of skin showing hyperkeratosis, epidermal necrosis, and intraepidermal
pustule formation.   The dermis is edematous and shows a marked acute inflammatory cell infiltration (hematoxylin and eosin stain H&E, ×132).

Fig. 11   (a) Overwintered prosomillet grains taken from an experimental field which caused severe cases of ATA.   Growth of F. sporotrichioides was heavy. (b) Grains of nontoxic normal prosomillet gathered in autumn 1943 from the experimental field.

Table 6   Toxicity of Overwintered Millet Grain Sampled at Different Seasons

Toxicity	Autumn		Winter		Spring	
	No. of samples	%	No. of samples	%	No. of samples	%
Toxic	16	20.5	7	11.2	38	100.0
Slightly toxic	18	23.1	19	30.1	—	—
Nontoxic	44	56.4	37	58.7	—	—
Total number of samples examined	78		63		38	

The contrast between the numbers of toxic samples collected during different seasons is brought out in a most striking fashion by an analysis of material relating to the years 1943 to 1944. Since that period was characterized by an exceedingly high degree of toxicity in cereals, there is good reason to regard the data from that year as most conclusive.

Let us consider the results obtained from one experimental plot in 1944 as an example. During the autumn-winter period, investigations were conducted on both the vegetative parts and grain of millet and on soil samples. Among the 102 samples of vegetative parts of millet, 48 were toxic or slightly toxic, and 54 samples were nontoxic. Of 36 samples of millet grain, only a single sample which had been gathered in January was toxic; slight toxicity was discerned in 10 cases, whereas 25 grain samples were shown to be entirely unaffected. Of three soil samples, one was found to be toxic.

The first spring samples were collected in April, from underneath the snow, from moist places free of snow, and from dry places. The investigation included eight samples of grain separated from the vegetative parts. In all, 16 extracts were prepared. Application of the extracts to rabbit skin produced edema, hemorrhage, and necrosis (Figs. 9 and 10). Two weeks later, after a full thaw, 11 samples were again collected from different parts of the plot. Strong toxicity was displayed in 22 extracts prepared from vegetative parts and from grain. Thus all the samples taken in spring proved to be highly toxic (Figs. 11a and 11b).

The findings from samples collected during the different seasons of the 1943/1944 crop are tabulated in Table 6. All the spring samples of millet were toxic, as compared to a much smaller proportion of toxicity among samples gathered in the autumn and winter.

3.10.5.1.2.5   Meteorological Conditions and the Incidence of ATA. A comprehensive analysis of the connection between various weather conditions prevailing in autumn and winter and the incidence of ATA in the 50 counties of the Orenburg district of the U.S.S.R. was carried out over the 7 years 1941/1942 to 1947/1948 (Table 7).

These dates show that the year of heaviest disease incidence, 1943/1944, was characterized by the following partially interrelated weather factors: (a) The temperatures in January and February were considerably higher than in most other years; (b) the depth of the snow cover in March far exceeded that of all other years; and (c) the depth at which the soil was frozen was less in autumn and was especially low in March.

The higher January-February temperatures, coupled with the heavy snow cover in March, evidently prevented the soil from freezing to its usual depth, which in ordinary years reaches 80 to 120 cm during February-March. As described in the section on conditions of toxin formation, temperatures in the -5 to $-10°C$ range and alternate freezing and thawing are exceedingly favorable to toxin production. The weather prevailing in 1943/1944 thus greatly favored accumulation of toxin in the cereals overwintering under the snow, and thus caused the disastrous outbreaks of ATA in 1944.

Table 7   Relations between Autumn and Winter Weather and the Incidence of ATA in the Orenburg District of the U.S.S.R., 1941/1942 to 1947/1948

Weather	Month	1941/1942	1942/1943	1943/1944	1944/1945	1945/1946	1946/1947	1947/1948
Temperature (°C)	January	-17.7	-18.8	-8.3	-16.2	-12.4	-16.6	-7.0
	February	-18.7	-15.7	-8.0	-20.8	-5.4	-15.1	-12.6
Snow cover (cm)	December	14	40	29	5	11	4	8
	January	22	53	50	5	29	6	15
	February	23	50	79	6	50	25	21
	March	25	43	108	17	70	18	28
Depth to which soil was frozen (cm)	November	52	21	12	25	40	33	n.r.
	December	n.r. ᵃ	36	14	30	50	75	41
	January	n.r.	n.r.	27	n.r.	50	106	68
	February	100	n.r.	29	80	n.r.	123	118
	March	n.r.	n.r.	10	80	91	n.r.	124

Incidence of ATA in 50 counties
Number of counties in which the population was affected by ATA at the following rates per 10,000 head:

		1941/1942	1942/1943	1943/1944	1944/1945	1945/1946	1946/1947	1947/1948
No disease		31	20	3	36	42	38	0
0-50 cases		12	17	3	14	8	11	0
50-500 cases		7	13	19	0	0	0	0
500-1000 cases		0	0	16	0	0	0	0
More than 1000 cases		0	0	9	0	0	0	0

ᵃn.r. = not recorded.

During the 30 years preceding the outbreaks of ATA in 1943/1944, similar combinations of low September-October rainfall (up to 25 mm) and relatively mild January-February temperatures (-9.5 to 12.5°C) were recorded in the Orenburg district only in 1924/1925 and 1934/1935. It is of interest to note that, according to Geminov [44, 45], cases of poisoning occurred in considerable numbers in the rural population of this district in those 2 years.

It has been shown in laboratory studies [70] that (a) toxin accumulation in cultures of fungi isolated from overwintered grain was intensified by fluctuations of temperature, such as those obtained by alternate freezing and thawing, and that (b) cultures of toxin producing species of F. poae and F. sporotrichioides grew well at temperatures as low as -2 to -7°C. Toxin formation was most active at temperatures just below zero, i.e., a few degrees above the growth minimum.

There were continuous years in which overwintered grains were ingested but no outbreak of ATA was recorded because the climatic and ecological conditions were not favorable for toxin production.

The conditions prevailing in spring 1944 appeared to have favored toxin formation by the relatively high temperatures and especially by the dense snow cover which prevented the soil from freezing at its usual depth. Alternate thawing and freezing occurred in the spring of that year more frequently than in other years, and this favored development of toxin-producing fungi on overwintering grain.

Although not substantiated by laboratory work, it is believed that the light rainfall in autumn 1943 may have been a contributory factor. The rain-free and relatively cold autumn nights, together with ample dew, may well have favored development of the toxin-producing fungi on vegetative cereal parts, from which they then proceeded to attack the grain under the favorable conditions of the following spring.

3.10.5.1.3   Comparison between the Toxicity of Fungal Cultures and Cereal Samples. It was interesting to compare toxicity of cultures with the toxicity of the cereal samples from which they had been isolated. Highly and mildly toxic fungi were detected in 1945 to 1949, both on toxic and on nontoxic overwintered cereals, whereas in 1944, when ATA was widely occurring in the Orenburg district, mildly and highly toxic cultures were isolated almost exclusively from toxic samples of overwintered cereals.

On comparing the type of reaction obtained from highly toxic fungi with that obtained from the respective cereal samples from which they had been isolated (samples gathered in spring 1944), it was noteworthy that toxic Cladosporium cultures, producing a reaction of the leukocytic edematous type, and Fusarium cultures, giving an edematous-hemorrhagic or necrotic reaction, had been isolated only from highly toxic cereal samples. Thus it is most probable that the decisive part in the development of toxicity in overwintered cereals was played by those Fusarium and Cladosporium species which gave these types of reaction.

Strukov and Mironov [188] studied the histology of the skin to which several toxic fungi from our material had been applied. These authors indicated that the tissue changes which developed following an application of ether extracts of these fungi to the skin of rabbits were similar to reactions caused by extracts from toxic millet.

As we have stated elsewhere [68] toxic species of Fusarium were always highly toxic when inoculated on sterilized grain.

Myasnikov [132], who tested F. poae in our laboratory, found that its toxic ingredient was similar to that contained in overwintered cereals. Similar results were obtained from our material by Olifson [138-140] who studied the chemical composition of millet after it had been experimentally infected with pure cultures of F. poae and F. sporotrichioides.

3.10.5.2   Biological Properties of Toxic Fusarium Fungi

The distribution of Fusarium fungi of the Sporotrichiella section is fairly wide-spread in plants, soils, and other substrates.

The remnants of vegetative parts of cereals in the field, grains which fall down at harvesting, and cereals which have been moved and heaped in the field and then wet by the rains, or harvested late in autumn after the rains have already started, all constitute a good medium for the development of fungi.

Since toxin may already be found in vegetative parts in the autumn [63, 71], the danger exists that the grains of cereals harvested in the spring will also be toxic.

The toxin is not equally distributed in the grain. In prosomillet grains more toxin was found in the glumes than in the grain itself. It was also observed that there are light and heavy grains. By separating them with 10 to 25% NaCl solution, it was found that the light grains which floated in the solution were toxic whereas the heavy grains were not toxic or less toxic [119, 120]. This observation led to the assumption that toxin does not invade the grain from without, but is produced within the grain. If the light and toxic grains are pressed slightly, they turn into powder, in contrast to the heavy nontoxic grains which cannot be easily ground. The powder derived from the grains infected with fungi contains the highest concentration of toxin.

The toxic fungus is believed to develop in the embryo of the grain and later the mycelium spreads through the whole grain. This is why the percentage of germination of overwintered grains infected with toxic fungi is much less than that of normal grains [74].

Climatic and ecological conditions are very important for the development of the fungus. If the winter is mild and temperatures not too low, the development of the Fusarium fungi is possible. The condition of the soil also influences the development of the fungi: the thicker the layer of snow, the less frozen the soil and the better the fungi develop. When the layer of snow is thinner and the soil freezes, fungi will not develop and toxin will not be produced.

From cultures of F. poae and F. sporotrichioides which had been isolated from overwintered cereals such as wheat, prosomillet, and oats, Bilai [19] found 11.5, 7.5, and 4.7%, respectively, to be toxic. In our isolates the incidence of toxicity was usually higher [68, 69, 72].

3.10.5.2.1   Temperature in Relation to the Stage of Fungus Development.
Pure cultures of F. poae and F. sporotrichioides, sown on liquid medium
(synthetic with starch and carbohydrate-peptone) and grown under different
temperature conditions, were assayed for toxicity at each of the three stages of
development, i.e., prior to sporification, at the time of abundant sporification,
and at the stage of senescence. Assays were carried out with ether extracts
of the liquid media and extracts of the fungal mass by skin tests on 22 rabbits.
The liquid substrates were passed through Seitz filters, and the sterile filtrates
obtained in this way were tested for toxicity on white mice by subcutaneous
injections of 0.2, 0.5, and 1.0 ml.

The results, previously reported by the present author [70, 73, 75],
showed that injection of filtrates of cultures obtained at different stages of
development at temperatures of 23 to 25 $^\circ$C did not cause death of white mice.

In the case of extracts of cultures maintained at low temperatures, as well
as those kept at 0 to +5 $^\circ$C with intervening freezing, death of the mice occurred
within 12 to 48 h, depending on the dosage and the fungal species. Death was
due to systemic toxic action. In every case, postmortem examination disclosed
necrosis in the digestive system and other organs. The highest toxicity was
displayed by filtrates of Fusarium obtained during the stage of abundant sporification from cultures grown at temperatures of -2 to -7 $^\circ$C and prior to
sporification from the -7 to -10 $^\circ$C series, whereas extracts obtained at an
advanced stage of senescence were considerably less toxic.

Application of liquid ether extracts of F. poae and F. sporotrichioides to
the skin of rabbits invariably produced a distinct skin reaction when the fungus
was at the sporification stage, even in the case of cultures grown at incubator
temperatures of 23 to 25 $^\circ$C. This circumstance seems to be connected with the
high concentration of the toxin concerned. However, it should be mentioned
that in every instance, cultures grown at low temperatures produced more
pronounced reactions than incubator-reared cultures. Highest toxicity was
associated with material obtained at the stage of abundant spore formation.

The data showed yet another interesting characteristic, namely that the
application of ether extracts of the liquid substrate to rabbit skin gave stronger
reaction than did extracts of the fungal film of F. poae and F. sporotrichioides
cultures. This indicates that the toxins of F. poae and F. sporotrichioides are
secreted into the surrounding medium, and thus act as exotoxins.

3.10.5.2.2   Effect of Substrate on the Growth of Toxic Fungi. We used
various vegetable parts and their grains (millet, wheat, barley, oats, rice,
potato, etc.) as well as solid and liquid synthetic media for culturing toxic fungi
[64, 68, 69, 77], and investigated numerous sources of nitrogenous and carbo-
hydrate nutrients and their effect on the formation of toxin by fungal cultures.
The following sources of nitrogen and carbon were tested: peptone, casein,
glycocoll, cystine, albumin, asparagine, tryptophan, tyrosine, glutamic acid,
urea, ammonium sulfate [$(NH_4)_2SO_4$], sodium nitrate ($NaNO_3$), sodium nitrite
($NaNO_2$), ammonium nitrate (NH_4NO_3), etc.; arabinose, dextrose, galactose,
glucose, saccharose, maltose, lactose; mannitol, starch, cellulose, sodium
citrate, sodium acetate, sodium oxalate. These tests involved 274 cultures of
toxic fungi.

The best nutrient sources among organic substances for F. poae and F. sporotrichioides proved to be carbohydrates (starch, glucose); peptone and asparagine were good suppliers of nitrogen. Best results among inorganic substances were obtained with ammonium sulfate [$(NH_4)_2SO_4$] and sodium nitrate ($NaNO_3$). Very meager growth of Fusarium was obtained with the use of organic acids. It is noteworthy that satisfactory development and production of toxic substances were obtained on filter paper with cultures of F. poae and F. sporotrichioides.

Certain problems relating to the nutritional physiology of F. sporotrichioides were investigated by Sarkisov and Kvashnina [171]. Among the substances tested, the best sources of nitrogen and carbon were found to be peptone, casein, asparagine, glucose, starch, and mannitol. Bilai [19] stated that aspartic acid, glutamic acid and its amides, alanine, glycocoll, ammonium carbonate, as well as gaseous ammonia, all provided suitable sources of nitrogen for Fusarium species of the Sporotrichiella section.

It may be assumed that toxin formation is also conditioned by the acidity of the medium. The most suitable pH values were found to be 4.6 to 5.4 for Fusarium [77].

3.10.5.2.3 Growth Characteristics Possibly Associated with Fungal Toxicity. Repeated investigations have shown that low temperatures promote rapid accumulation of toxin in cultures of F. poae and F. sporotrichioides, despite the slow growth of the mycelia. Cultures grown at high temperatures, although producing luxuriant mycelial growth, were nontoxic or only slightly toxic.

It is of interest to point out that decrease of sporification was occasionally observed in cultures of F. poae and F. sporotrichioides grown at 24 to 28°C. Such cultures, when subjected to rabbit skin tests, were shown to be nontoxic or slightly toxic. Cultures of the toxic fungi F. poae and F. sporotrichioides, grown at low temperature or under conditions of alternating freezing and thawing, were characterized by prolific production of spores and high toxicity.

The presence of abundant nonsporifying aerial mycelium usually coincided with the absence of toxicity. On the other hand, scanty aerial mycelium with a large number of spores would normally be associated with high toxicity.

Toxicity of these cultures is apparently associated with intensive production of spores.

3.10.6   Clinical Characteristics of
          Alimentary Toxic Aleukia

Many different names have been used for the description of the clinical aspects of ATA: septic angina [26-28, 45, 46, 82, 96, 97, 100, 101, 107, 134, 135, 152, 180, 184, 208], alimentary hemorrhagic aleukia [131, 192], alimentary aleukia [30], agranulocytosis [85, 105], cereal agranulocytosis [58], acute myelotoxicosis [108], alimentary panhematopathy [210], aplastic mesenchymopathy [85], and endemic panmyelotoxicosis [114, 115].

A Committee of the Soviet Health Ministry (1943) decided upon alimentary toxic aleukia as the proper official name for the disease. This name emphasized the progressive leukopenia and the fact that ingestion of grain (alimentary) and the secretion of toxin are necessary for an outbreak of this disease.

In order to establish a diagnosis of ATA, the recent dietary history of the patients had to be carefully investigated. The quantity and duration of feeding on overwintered toxic grains or their products had to be determined.

The clinical findings were described by Chilikin [26-28], Lyasas [105], Manburg [108], Manburg and Rachalskij [109], Myasnikov [131], Nesterov [135], Romanova [154], and Yefremov [207].

3.10.6.1   Stages of Development
            of ATA

The clinical features of ATA are usually divided into four stages. If the disease is diagnosed during the first stage, and even at the transition from the second to third stages, early hospitalization may still enable the patient's life to be saved. If, however, the disease is only detected during the third stage, the patient's condition is usually desperate and in most cases death cannot be prevented. Only very few patients in the third stage survive.

3.10.6.1.1   The First Stage. The characteristic symptoms of this stage appear a short time after ingestion of the toxic grains. Sometimes they appear after a single meal of overwintered toxic grains and disappear completely even if the patient continues to eat the grains. The characteristics of the first stage include primary changes, with local symptoms, in the buccal cavity and gastrointestinal tract. Shortly after eating food prepared from toxic grain, the patient feels a burning sensation in the mouth, tongue, throat, palate, esophagus, and stomach as a result of the action of the toxin on the mucous membranes. The tongue may feel swollen and stiff, and the mucosa of the oral cavity may be hyperemic. After 3 to 4 days a mild gingivitis, stomatitis, glositis, and esophagitis develops, and inflammation of the gastric and intestinal mucosa results in vomiting, diarrhea, and abdominal pain [27, 44, 45]. In most cases excessive salivation, headache, dizziness, weakness, fatigue, and tachycardia accompany this stage, and there may be fever and sweating. Usually the temperature of the body does not rise and stays normal. The leukocyte count may already decrease in this stage to levels of 2000 mm^{-3}, with relative lymphocytosis, and there may be an increased erythrocyte sedimentation rate [207].

The danger exists that this stage may not always be detected because it appears and disappears relatively quickly; the patient may become accustomed to the toxin and a quiescent period follows while the effects of the toxin accumulate and the patient enters the second stage. The first stage may last from 3 to 9 days [28, 135].

3.10.6.1.2   The Second Stage. This stage is often called the latent stage
or incubation period [28] because the patient feels well and is capable of
normal activity [206, 207]; it is also called the leukopenic stage [109, 154]
because its main features are disturbances in the bone marrow and the hematopoietic system, characterized by a progressive leukopenia, a granulopenia,
and a relative lymphocytosis. In addition, there is anemia and a decrease in
the erythrocytes in the platelet count and hemoglobin.

The progressive decrease in leukocytes lowers the resistance of the body
to bacterial infection [28]. In addition to changes in the hematopoietic
system, there are also disturbances in the central and autonomic nervous
systems. Weakness, vertigo, fatigue, headache, palpitations, and mild
asthmatic symptoms may occur. The skin and mucous membranes may be
icteric, the pupils dilated, the pulse soft and labile, and the blood pressure
decreased. The body temperature does not exceed 38°C and the patient may
even be afebrile. There may be diarrhea or constipation. In this stage visible
hemorrhagic spots appear on the skin; the appearance of hemorrhages marks
the transition from the second to the third stage of the disease.

The normal duration of this stage is usually from 3 to 4 weeks, but it may
extend over a period of 2 to 8 weeks. If consumption of toxic grain continues,
the symptoms of the third stage rapidly develop. According to Chilikin [28]
the hemorrhagic syndrome at the end of the second stage develops into
necrotic angina.

3.10.6.1.3   The Third Stage. The transition from the second to the third
stage is sudden. At this stage the patient's resistance is already low, and
violent symptoms may be present, especially under the influence of stress,
associated with physical exertion and fatigue.

The first visible sign of this stage is the appearance of petechial hemorrhages on the skin of the trunk, in the axillary and inguinal areas, on the
lateral surfaces of the arms and thighs, on the chest (Figs. 12 and 13), and, in
serious cases, on the face and head. The petechial hemorrhages vary from a
few millimeters to a few centimeters in diameter [27, 28].

As a result of increased capillary fragility, any slight trauma may cause
the hemorrhages to increase in size. Hemorrhages may also be found on the
mucous membranes of the mouth and tongue, and on the soft palate and tonsils.
Nasal, gastric, and intestinal hemorrhages and hemorrhagic diathesis may
occur [28, 207].

Necrotic angina begins in the form of catarrhal symptoms and necrotic
changes soon appear in the mouth, throat, and esophagus with difficulty and
pain on swallowing. The necrotic lesions may extend to the uvula, gums,
buccal mucosa, larynx, vocal cords, lungs, stomach, and intestines (small
and large) and to other internal organs [28], and are usually contaminated with
a variety of avirulent bacteria. The necrotic areas are an excellent medium
for bacterial infection because of the lowered resistance of the body due to the
damage to the hematopoietic and reticuloendothelial systems. Bacterial infection causes an unpleasant odor from the mouth due to the enzymatic activity of
the bacteria on proteins. Areas of necrosis may also appear on the lips and on
the skin of the fingers, nose, jaws, and eyes (Fig. 14).

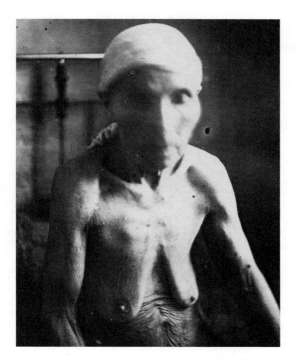

Fig. 12   Petechial spots, first small and red, later blue or dark, caused by intradermal or submucous hemorrhage on the chest and on the left arm.

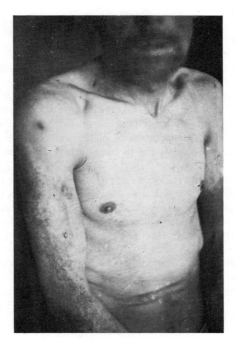

Fig. 13   Hemorrhage on the right and left arms and initially in the chest.

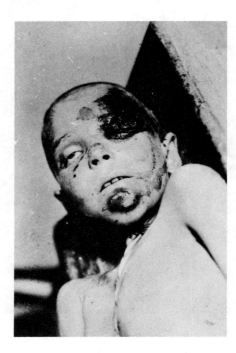

Fig. 14   Necrotic lesions around the eye and on the face in a child who died
from ATA disease.

The regional lymph nodes are frequently enlarged.   The submandibular and
cervical lymph nodes may become so large, and the adjoining connective tissue
so edematous, that the patient experiences difficulty in opening his mouth.
Esophageal lesions may occur and involvement of the epiglottis may cause
laryngeal edema and aphonia (loss of voice).   In such cases, death may occur
by strangulation.   Death of about 30% and sometimes a higher percentage of the
patients was directly related to stenosis of the glottis [148].   The acute
necrotic changes were associated with the septic state characterized by
extreme weakness and apathy.

The blood abnormalities observed initially in the first and second stages
become intensified during the third stage.   The leukopenia increases to counts
of 100 mm^{-3} or even fewer leukocytes per cubic millimeter.   The lymphocytes
may constitute 90% of the white cells present, the number of thrombocytes
decreases below 5000 mm^{-3}, and the erythrocytes below 1 million mm^{-3}.

The blood sedimentation rate is increased (and in most cases shows a
deficiency of prothrombin).   The prothrombin time ranges between 20 and 56
sec, and the clotting time is usually not much prolonged.   There may be a
deficiency in fibrinogen in severe cases [47].

Some investigators found that patients suffer an acute parenchymatous hepatitis accompanied by jaundice. Bronchopneumonia, pulmonary hemorrhages, and lung abscesses are frequent complications.

3.10.6.1.4   The Fourth Stage. This is the stage of convalescence, the course and duration of which depend on the intensity of the toxicosis. Therefore the duration of the recovery period is variable. Only about 3 to 4 weeks of treatment, or sometimes longer, is needed for the disappearance of necrotic lesions and hemorrhagic diathesis and also the bacterial infections. Usually 2 months or more elapse before the blood-forming capacity of the bone marrow returns to normal: as a general rule, first the leukocytes, then the granulocytes, the platelets, and subsequently the erythrocytes.

3.10.7   Pathology of Organs in Man
          and Animals

Toxic materials from overwintered grains had local and general effects on the tissues of man and animals.

3.10.7.1   Effect of F. poae and F.
            sporotrichioides in
            Overwintered Grain on
            Various Organs in Man

The local action was manifested by clinical burning sensations in the mouth, soft palate, and tongue. These phenomena usually passed when the patient put an end to eating products made of toxic grains. If he continued to consume the fatal grains, then the first signs of toxicosis appeared in the pharynx, esophagus [46], and stomach [108], and later various hemorrhagic syndromes appeared, characterized by hemorrhagic rash on the skin of the chest (Fig. 15), trunk, abdomen, legs, and arms, and even on the face. The hemorrhagic petechiae appeared most abundantly at the start of the development of necrotic angina [28, 94, 109]. Hemorrhagic and necrotic lesions developed also in the stomach and intestines, including all of the digestive tract, causing serious changes in intestinal function [28, 30, 108, 109]. Hemorrhage and necrosis of the appendix and cecum, and inflammation of the rectum also developed [28]. Hemorrhages were also present in different visceral organs, in the adrenal and the thyroid glands, gonads, uterus, and chiefly in the pleura. Pulmonary changes of bronchopneumonia and severe hemorrhages in the lung tissue with pneumonic abscesses frequently developed [28, 135]. Marked changes were observed in the heart accompanied by vascular insufficiency and arterial hypotonia. Pressure was very low, and sometimes thrombophlebitis and endocarditis were indicated [135]. These changes depended on the serious inflammation of the blood vessels and on the condition of the endocrine system. In the third stage of the disease hemorrhages in the liver were revealed with

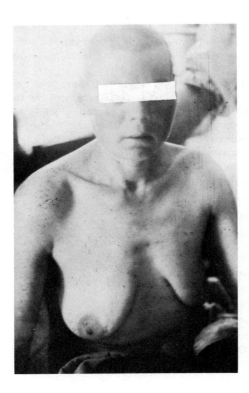

Fig. 15   Hemorrhagic rash on the skin of the chest and arms.

acute parenchymatous hepatitis accompanied by jaundice [28, 135, 154, 206, 207], and various changes in the glucose, protein, mineral, and other metabolisms of the liver were observed [28, 108]. Hemorrhages and necrosis in the kidneys appeared as well (Fig. 16). In the necrotic angina stage, severe changes were observed in the lymph nodes which became edematous and were characterized by disappearance of the lymphoid formation. The entire lymphatic and reticuloendothelial systems were affected and showed proliferation of the red blood cells and of the endothelial capillaries and sinuses [207]. Various investigators have observed severe changes in the central and autonomic nervous systems by ATA [60, 86, 150], such as impaired nervous reflexes, meningism, general depression and hyperesthesia, encephalitis, cerebral hemorrhages, and destructive lesions in nervous and sympathetic ganglia [60, 178, 179].

According to Tomina [196], toxins from overwintered cereals which were absorbed in the stomach and intestines had a cumulative effect in various organs and tissues.

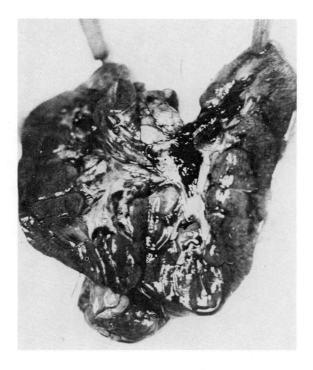

Fig. 16   Kidney of fatal case of ATA showing marked hemorrhage in the pelvic mucosa.

The most severe effects were on the hematopoietic system which correlated with the development of the different stages of the disease and resulted in depression of leukopoiesis, erythropoiesis, and thrombopoiesis [84, 87, 194]. Progressive leukopenia appeared (the leukocyte count dropped to 100 mm^{-3} or even lower) as well as lymphocytosis (to 90%) and a decrease in erythrocytes (which dropped to 1 million mm^{-3}). The hemoglobin content dropped to 8% and the granulocytes completely disappeared; at the same time, the sedimentation rate increased and showed a deficiency of prothrombin as a result of the irritation of the bone marrow [26-28, 84]. Frequently, in severe cases, destructive and hemorrhagic lesions appeared in the blood circulation which caused thrombosis in the blood vessels of different organs.

The destruction, devastation, and sometimes atrophy of the bone marrow [28, 30, 205-207] showed serious changes in the organism after consumption over a long period of time of overwintered grain contaminated by the toxic strains of F. poae and F. sporotrichioides. Hemorrhagic diathesis, necrotic angina, sepsis, and severe hematological changes developed as well [28, 30, 56, 108, 109, 207].

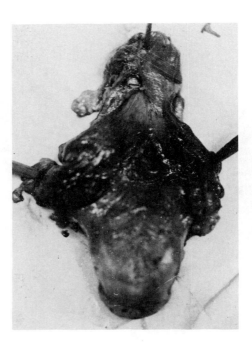

Fig. 17   Necrosis of pharyngeal mucosa caused by F. poae which was isolated from prosomillet wintered under snow cover. The toxic prosomillet grain was eaten by this patient.

An important contribution to the pathogenesis of ATA was that of Strukov [187] and Strukov and Tishchenko [186, 189], who showed that disturbances of the hematopoietic system were reversible and did not lead to the destruction of bone marrow. Strukov [187] and Aleshin et al. [2, 3] thought that toxins of overwintered cereals did not act primarily on the bone marrow but on an extramedullary apparatus which regulated the hematopoietic, autonomic nervous, and endocrine systems.

The necrotic changes in the final stage of ATA developed initially in the pharyngeal tonsils, and later in so-called necrotic angina, in the throat, and even in the esophagus; severe gangrenous pharyngitis occurred as well (Fig. 17). The necrotic angina in the throat brought about an increase of fever. In severe cases signs of glottis edema with asphyxia were observed [148, 184]. The necrotic appearance caused weakness, apathy, and damage of the leukocytic, phagocytic, and also reticuloendothelial functions of the organism. Necrotic lesions were present along the entire gastrointestinal tract and also in other organs.

The strong and profuse menstrual and nose bleeding (Fig. 18) were fatal in many cases.

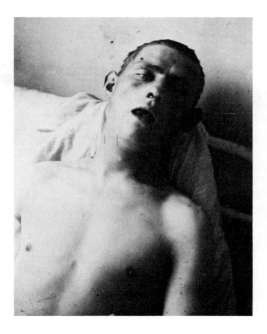

Fig. 18   General views of a patient with a severe form of ATA: nosebleed (epistaxis), respiratory distress, hemorrhage on the left arm.

Recovery depended on the amount of consumed toxic grain, on the type of therapy, and chiefly on the presence or absence of additional complications.

Sometimes the disease recurred with hemorrhagic and necrotic syndromes after resuming consumption of infected grain, or after physical strain [27, 28, 135].

3.10.7.2   In Animals

Some types of experiments have been carried out to study the effect of over-wintered toxic grains and grains infected with F. poae and F. sporotrichioides strains on different laboratory animals, by oral feeding, parenteral tests, and also by skin tests on rabbits. Skin reactions, caused by the action of toxic overwintered grains and by toxic Fusarium extracts, were distinguished both by their external appearance (application on skin) and histological changes (Figs. 19-21). The toxins of F. poae and F. sporotrichioides had, thus, both a localized and generalized toxic effect. The localized effect was first apparent as an inflammatory reaction and was accompanied mainly by subsequent skin hemorrhage or necrosis at the site of the toxin application. The general effect was apparent in defective blood production, acute degenerative processes in the

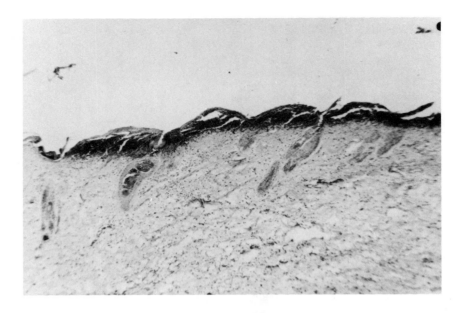

Fig. 19   F. poae grown at +12°C. Total necrosis of the epidermis with leukocyte infiltration.   The epidermis is presented in only one to two layers of
necrotic cells. Necrosis of hair follicles (H&E, ✕102).

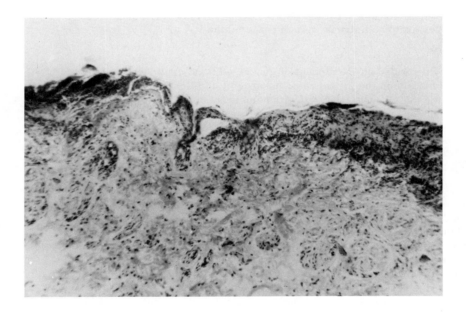

Fig. 20   F. sporotrichioides grown at +4°C. Necrosis of the epidermis with
an infiltration of leukocytes in the upper layers of the dermis. Here necrosis
is found accompanied by edema and leukocyte infiltration (H&E, ✕135).

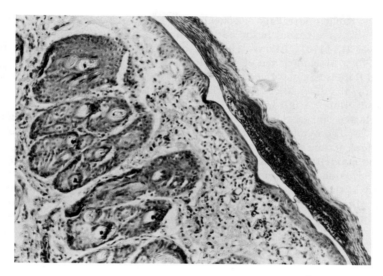

Fig. 21   F. sporotrichioides var. tricinctum grown at +12°C.   A thick layer
of parakeratotic epithelium heavily infiltrated with polymorphonuclear leukocytes is seen on the surface.   The dermis shows a moderate inflammatory cell
infiltration (H&E, ×132).

internal organs, and extreme hyperemia and hemorrhages, especially of the
digestive tract.   Pentman [147] used the skin test on rabbits, which was not
used again until our laboratory [29, 68, 69, 78, 79, 118, 122-128] and the All-
Union Research Laboratory of Toxic Fungi of U.S.S.R. Agriculture Ministry
[166-169, 171] carried out large numbers of skin tests in 1942.   Today skin
tests are considered a reliable method for determining the toxicity of overwintered grains and various strains of toxic Fusarium.

In the same way that the clinical syndrome of ATA varies from patient to
patient, so do the symptoms in animals vary according to the potency and
quality of the toxin, the means of administration, and the sensitivity of the
animal.   These variations led Bilai [19] to refer to the polymorphism of the
disease.   For many years the typical clinical syndrome of ATA in man could
not be produced in laboratory animals.   When toxic overwintered grains, as
well as normal grains and other different solid and liquid media infected by
toxic F. poae and F. sporotrichioides, were fed to mice, rats, guinea pigs,
rabbits, dogs, horses, chickens, pigs, sheep, and cattle, or toxic extracts
were pipetted orally or parenterally, the animals developed clinical syndromes
which differed from the typical picture of ATA in man [8, 9, 19, 68-75, 102-
104, 106, 155, 160, 162, 167, 169].

The clinical and pathological findings revealed stomatitis, gingivitis,
glositis, esophagitis, hyperemia, hemorrhages and necrosis of the gastrointestinal tract, and even hemorrhagic diathesis and degeneration in the liver,
kidneys, heart, adrenal and thyroid glands, and nervous system, which

sometimes caused paralysis of convulsion. However, characteristic changes
were not always apparent in the cellular blood picture. Sarkisov [168],
Sarkisov et al. [170], Alisova [7], Alisova and Mironov [4, 6], and Joffe
[64, 68, 73, 75] studied the effect of toxic F. poae and F. sporotrichioides on
cats, and Rubinstein and Lyass [162] on cats and monkeys, and succeeded in
reproducing the disease of ATA. The death of the cats followed a failure in
blood production. A fall in hemoglobin, red cells, and leukocytes was observed
with a relative increase in lymphocytes. The lowest leukocyte count found was
100 mm^{-3}. The symptoms included vomiting, a hemorrhagic diathesis, and
neurological disturbances.

In the majority of cases, the cats died on the 6th to 12th day [68, 71].
Autopsy revealed marked hyperemia and hemorrhages of internal organs,
especially of the digestive tract, liver, and kidneys, and extreme changes in
the adrenal glands. Histopathological examinations of organs of cats dying after
infection with F. poae and F. sporotrichioides and hematological analysis
revealed changes in the tissues and in the blood-component elements which
were similar to those seen in ATA in man [201].

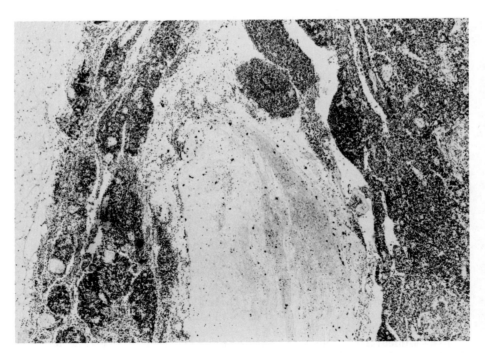

Fig. 22   Retroperitoneal lymph node. Mouse dying 1 day after a subcutaneous
dose of F. sporotrichioides extract: cystic distension and hemorrhage
(H&E, ×40).

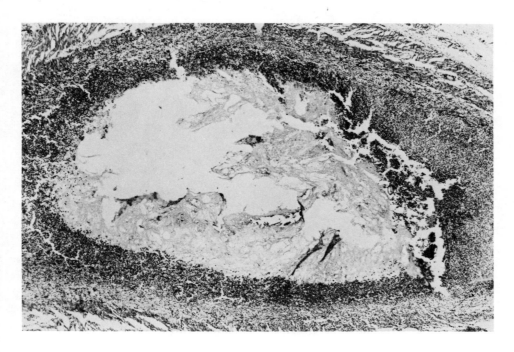

Fig. 23   Esophagus. Rat killed in extremis 15 days after the first, and 2
days after the last, of 5 doses of F. poae extract, given by stomach tube:
local distension and blockage with cellular debris (H&E, ×40).

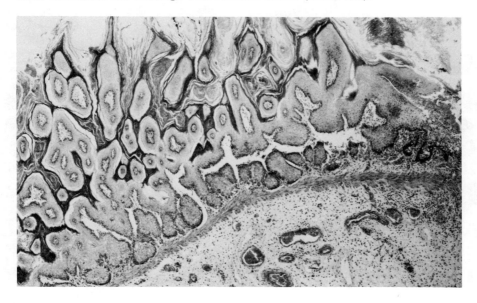

Fig. 24   Forestomach. Female rat dying 23 days after the first, and 1 day
after the last (doubled), of 11 doses of F. poae extract: squamous cell hyperplasia, edema of the stomach wall, and (on the right) ulceration (H&E, ×100).

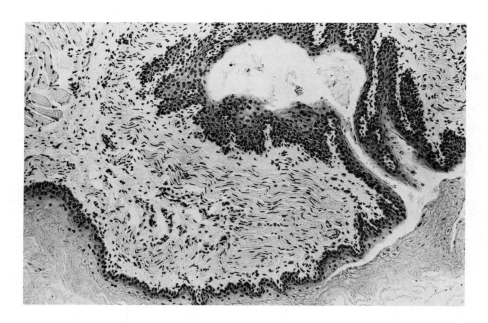

Fig. 25   Esophagus.   Rat killed 9 weeks after the first, and 7 days after the
last, of 10 doses of F. poae extract given by stomach tube: invagination and
basal cell hyperplasia of the squamous epithelium (H&E, ×100).

Fig. 26   Skin.   Mouse dying 10 weeks after the beginning of approximately
weekly application of F. poae extract: areas of desquamation of the keratinized
epithelium, hyperplasia of the basal cells, and foci of infection, some sur-
rounded by inflammatory cells (H&E, ×70).

Schoental and Joffe [172] suggested that the crude extract obtained from toxic cultures of F. poae and F. sporotrichioides have two types of action, local and systemic, in rats and mice. We have used the extracts from these Fusarium species to test their long-term effects in rodents when administered in various ways. Single doses, on the order of 0.1 ml, were toxic to young animals, who died within a few days. Smaller doses allowed the animals to survive longer, and we were able to repeat the dosage at various time intervals. The authors wished to draw attention to the chronic lesions that developed in animals which had survived for several weeks or months and which suggested that the Fusarium metabolites may have been carcinogenic and immunosuppressive. Local applications of nonlethal doses of the F. poae and F. sporotrichioides extracts to the skin or to the esophagus and stomach, caused irritation, edema, and ulceration, followed by keratinization, desquamation, and healing, accompanied by basal cell hyperplasia of the squamous epithelium. Immunosuppression and induction of hyperplasia suggested that the extracts may have contained carcinogenic constituents. This appearance is clearly shown in Figs. 22, 23, 24, 25a and 26. The use of moldy cereals, contaminated by the inflammatory and irritant metabolites of F. poae and F. sporotrichioides, may have played a part in the development of tumors of the digestive tract.

The larger doses of crude extract caused death within 1 day or longer regardless of the route of administration. The main lesions were present in the lymphoid tissues, lung, heart, kidney, and digestive tract. Figure 22 shows the cystic distension and hemorrhage in the retroperitoneal lymph node of mice after a subcutaneous dose of F. sporotrichioides extract. Some of the more acute effects induced by our crude Fusarium extracts resembled those described in animals treated with certain trichothecenes [14, 112, 165, 191].

The trichothecenes were tetracyclic sesquiterpenoids [51] which contained a double bond and a spiroepoxy ring at carbons 12 and 13, essential for biological activity [13]. The effects of esterification, oxidation, and so on, have still to be explored in detail. Only one long-term study appears to have been published [111]. These authors did not find liver tumors in rats and trout given T-2 toxin from F. tricinctum in their diet for 8 months. However, the epoxide-containing trichothecenes would be expected not to require metabolic transformation in the liver and are likely to act mainly at the site of application. Our results are compatible with such action and suggest that neoplastic lesions may develop among the animals which are still surviving.

Akhmeteli et al. [1] carried out a trial on mice with extracts of barley infected by F. sporotrichioides and also discovered the blastogenic properties. In contradiction to these results, Lindenfelser et al. [99] carried out extensive studies on the biological effects of trichothecene compounds, particularly T-2 toxin and diacetoxyscirpenol (Sec. 2.5.1.3), on mouse tissue and revealed that these substances caused inflammation and tissue necrosis when applied to the skin of rats and mice, but did not cause papillomas.

The trichothecene compounds, diacetoxyscirpenol, T-2 toxin, and HT toxin, caused an inflammatory skin response on rabbits, guinea pigs, and albino rats [10, 13, 50] but were not carcinogenic.

3.10.8   Prophylaxis and Treatment

The most important prophylactic measure is to refrain from eating over-wintered toxic grains. The primary preventive measure, therefore, consists in educating the rural population as to the etiology and clinical symptoms of ATA; such measures have greatly reduced outbreaks of ATA. When the out-breaks of the disease were first reported, medical teams were sent to the affected areas and the population was examined clinically and hematologically.

Grain samples should be examined for toxicity by skin tests. At the same time, toxic cereals should be replaced by normal grain. When ATA is detected in the second stage, treatment should include blood transfusion and administration of nucleic acid and calcium preparations, sulfonamides, and vitamins C and K [42, 98]. When the number of leukocytes declines below 3000 mm^{-3} hospitalization is recommended. The measures employed in the second stage are also indicated in the third stage, but treatment is more intense. Following recovery, a rich diet is given for 1 month and the patient remains under periodic hematologic checkup [98].

3.10.9   Summary

The studies reported here were carried out in the U.S.S.R. in connection with the author's work on alimentary toxic aleukia.

The disease is caused by the fungi F. poae and F. sporotrichioides, which are the principal factors of the fatal human toxicosis. ATA occurred in various districts, republics, regions, and counties in the Soviet Union and especially in the Orenburg district where the author worked for 8 years.

Mild winters with abundant snow, followed by frequent alternate freezing and thawing in the spring, were favorable conditions for the development of the toxic Fusarium fungi. The rural people who consumed overwintered and snow covered grains infected by toxic fungi of F. poae and F. sporotrichioides were heavily affected. The first outbreaks of ATA usually appeared at the end of April or May and sometimes sporadic cases occurred during different times of the year. The disease developed in any individual, irrespective of age or sex, who consumed products from toxic overwintered grains.

The toxicity of overwintered grain and Fusarium fungi was studied by an application of their extracts to the skin of rabbits, which caused hyperemia, necrosis, and other severe reactions such as edema and hemorrhages, and also by feeding Fusarium toxins to laboratory animals. F. poae and F. sporo-trichioides when fed to cats produced the same fatal ATA syndrome that appeared in man.

The toxins of F. poae and F. sporotrichioides were identified by the Russian scientists as steroids, and according to U.S. researchers F. tri-cinctum was found to be a diacetoxyscripenol and T-2 toxin. These toxins are sesquiterpenoids and derivatives of the trichothecene group of toxins, different from those described by Russian investigators. F. tricinctum was one of the fungi most frequently isolated from corn associated only with toxicity in domestic animals in Wisconsin. We suggested that the toxin structures of

F. poae and F. sporotrichioides isolated from overwintered grains were not
yet finally established.

In many recent studies in the United States on the toxic effects of fungi of
the Sporotrichiella section on animals, reference has been made to the toxicosis, termed alimentary toxic aleukia (ATA), caused by F. poae and F.
sporotrichioides in man in the U.S.S.R. Attempts have been made to relate
the symptoms observed in various animals by F. tricinctum toxins, to published descriptions of ATA. Such comparisons are, in our view, premature
at the present stage of knowledge. The properties of isolates from overwintered grain in the U.S.S.R., which caused ATA in humans ingesting such
cereals, cannot properly be related to the properties of F. tricinctum isolates
causing toxic symptoms in animals in other countries, until and unless isolates
have been grown and tested under very similar conditions. Therefore we
decided to study carefully the toxicity of authentic strains of F. poae and F.
sporotrichioides isolated from overwintered grain, which were responsible for
the disease of ATA. In addition we arranged the classification of the toxic
strains of Fusarium of the Sporotrichiella section, which was very important
for an advanced discussion of the mycotoxin problem. The existing taxonomy
caused much confusion in classifying this group of fungi [113, 185]. The
majority of strains isolated from overwintered grains belonged to toxic species
of F. poae and F. sporotrichioides, which were characterized by cryophilic
properties. Temperature fluctuations (freezing and thawing) and abundant
spore formation, when grown at low temperatures, brought about a very intensive accumulation of toxin.

The clinical signs and symptoms and the pathological findings of ATA were
divided into four stages associated with destructive changes in the bone marrow
and in the hematopoietic system, hemorrhagic and necrotic lesions of the
mouth, throat (septic angina), and gastrointestinal tract, and also disturbances,
degenerations, and complications in the internal tissues and organs. In
addition the lymphatic and reticuloendothelial systems showed reticulosis.

The clinical symptoms in animals varied according to the potency and
quality of the toxin, the route of administration, and the sensitivity of the
animal.

Crude extracts of smaller doses of F. poae and F. sporotrichioides
developed chronic lesions, causing carcinogenic and immunosuppressive
properties in mice and rats after application of the Fusarium extracts to the
skin, esophagus, and stomach. Large doses caused depletion of the lymphoid
tissues and of various other organs, and killed the animals within 1 day or
more.

ATA was successfully treated by elimination of the toxic grains from the
diet as well as administration of various medicines and blood transfusions.

3.10.10   Recent Developments

The role of the toxic F. poae and F. sporotrichioides as etiological agents in
ATA and the symptoms of this disease in men have been described.

The toxins obtained from F. poae and F. sporotrichioides associated with
ATA were identified in the 1940s by Russian researchers as steroids [139-
146]. Research done in Japan [198] and in the United States [116] has
recently established that F. sporotrichioides NRRL 3510 and poaefusarin
obtained from F. poae (provided by Bilai, Kiev, Ukr. S.S.R.) which were
associated with ATA, produced trichothecene compounds such as neosolaniol,
HT-2 toxin, butenolide, T-2 tetraol, zearalenone, and chiefly T-2 toxin. The
presence of steroid-type compounds in these samples could not be confirmed.
In view of the continuing controversy we again engaged with Dr. B. Yagen in a
much needed study with our original and authentic strains of F. poae and F.
sporotrichioides isolated from overwintered grain collected at the time of the
fatal ATA outbreaks in the province of Orenburg in the Soviet Union. We have
examined the toxic metabolites available in our laboratory from the 131
isolates (106 F. sporotrichioides and 25 F. poae) cultivated at low temperatures. These toxic Fusarium species were obtained from monoconidial
cultures and maintained on sterilized soil and on standard potato-dextrose-
agar (PDA) medium at 3°C. The cultures were inoculated and grown on wheat
or millet. The identification of the isolated compounds and chiefly of T-2
toxin was determined by thin-layer chromatography (TLC), gas-liquid chromatography (GLC), spectroscopic analyses, and also by bioassay of rabbit skin
tests [212].

This is the first screening report on the distribution of T-2 toxin-
producing Fusarium fungi isolated in the U.S.S.R., and associated with ATA.
Our study has shown that more than 95% of the F. poae and F. sporotrichioides
isolates produced T-2 toxin in various quantities. Among the isolates in our
collection there were many which produced gram quantities of T-2 toxin when
grown on 1 kg wheat or millet at 12 or 5°C for a period of 21 and 45 days,
respectively. Those were isolates involved in the more severe cases of ATA
in the Orenburg district.

A good correlation was demonstrated between T-2 toxin detection by TLC
and inflammatory skin reactions of rabbits, and the comparison of amounts of
T-2 toxin determined by GLC also corresponded to the rabbit skin response.

We also undertook a comparative study of the amount of T-2 toxin produced by F. poae and F. sporotrichioides and F. sporotrichioides var.
tricinctum obtained from overwintered cereal grains involved in ATA disease
in men with the yield of this toxin from different sources [213]. These
isolates were obtained from fescue hay, corn, and wheat in various areas of
the United States which were associated with outbreaks of severe toxicity in
farm animals. These isolates have been referred to by American and Japanese
investigators mostly as F. tricinctum [80]. Closer study has shown them to
belong to various species of the Sporotrichiella section, especially F. poae and
F. sporotrichioides [75, 76, 80].

There were, in general, well-defined differences in the yield of T-2 toxin
between the strains isolated from overwintered grain sources in the U.S.S.R.
and strains isolated in the United States. The results are summarized in
Table 8.

Table 8 The Toxic Strains Used, Their Origin, and T-2 Toxin Yield

Strain	Origin	Supplied by	Rabbit skin bioassay[a]	TLC assay[a]	Yield of T-2 toxin (mg/10 g wheat grain)[b]
		Fusarium poae			
60/9	Millet	U.S.S.R.	A. Z. Joffe[c]	++++	Very strong	20.0
396	Millet	U.S.S.R.	A. Z. Joffe	++++	Very strong	21.0
792	Barley	U.S.S.R.	A. Z. Joffe	+++	Strong	6.2
958	Wheat	U.S.S.R.	A. Z. Joffe	+++	Strong	8.0
NRRL 3287	Unknown	U.S.A.	C. W. Hesseltine[d]	+	Slight	1.0
NRRL 3299	Corn	U.S.A.	C. W. Hesseltine	+++	Strong	5.8
T-2	Corn	France	W. F. O. Marasas[f]	++	Medium	3.4
		F. sporotrichioides			
60/10	Millet	U.S.S.R.	A. Z. Joffe	++	Strong	10.3
347	Millet	U.S.S.R.	A. Z. Joffe	++++	Very strong	23.2
351	Millet	U.S.S.R.	A. Z. Joffe	+++	Very strong	15.2
738	Millet	U.S.S.R.	A. Z. Joffe	+++	Strong	7.8
921	Rye	U.S.S.R.	A. Z. Joffe	++++	Very strong	24.0
1182	Wheat	U.S.S.R.	A. Z. Joffe	+++	Strong	10.2
1823	Barley	U.S.S.R.	A. Z. Joffe	+++	Strong	4.6
NRRL 3249	Tall fescue	U.S.A.	C. W. Hesseltine	++	Medium	4.0
NRRL 5908	Tall fescue	U.S.A.	H. R. Burmeister[d]	+	Slight	0.6
2061-C	Corn cobs	U.S.A.	C. J. Mirocha[e]	+	Slight	1.5
YN-13	Corn	U.S.A.	C. J. Mirocha	+	Slight	2.0
		F. sporotrichioides var. *tricinctum*			
NRRL 3509	Unknown	U.S.A.	C. W. Hesseltine	+	Slight	0.5

[a]The intensity of gray-brown spot at RF of T-2 toxin. [b]As determined by GLC. [c]Department of Botany, The Hebrew University of Jerusalem, Israel. [d]Northern Regional Research Lab., USDA, Peoria, Illinois 61604. [e]Department of Plant Pathology, University of Minnesota, St. Paul, Minnesota 55101. [f]Plant Protection Research Institute, Pretoria, South Africa.

The table shows that the U.S.S.R. strains produced several times as much toxin as the others, which proves the fundamental influence of background conditions on the capacity of T-2 toxin production. The special conditions under which overwintering grain was then affected by the Fusarium species have been described in detail by Joffe [65, 68-75]. These conditions might have been conducive to the extraordinarily high levels of T-2 toxin formation on wheat, millet, rye, oats, barley, and buckwheat. The U.S. strains producing T-2 toxin have not been isolated from substrates subjected to such extreme conditions, and their level of toxicity has remained much lower.

The importance of the origin of toxigenic strains is clearly shown by our comparison of T-2 toxin production by authentic ATA strains from the Soviet Union and strains used in research in the United States. It is worthy of special attention that strains associated with ATA were analyzed 30 years after they had first been isolated. This indicates remarkable preservation of toxigenic capability for high-level production of T-2 toxin in these strains.

References

1. M. A. Akhmeteli, A. B. Linnik, K. S. Cernov, V. M. Voronin, A. Ja. Hesina, N. A. Guseva, and L. M. Sabad (1973): Study of toxins isolated from grain infected with Fusarium sporotrichioides. Pure Appl. Chem. 35:209-215.

2. B. Aleshin and E. Eyngorn (1944): Changes in blood by septic angina and experimental agranulocytosis, in: Data on Septic Angina. Proc. First Kharkov Med. Inst. a. Chkalov Inst. Epidemiol. Microbiol. 1:135-172. Chkalov.

3. B. V. Aleshin, Sh. A. Burstein, and B. I. Chernyak (1947): Hematopoietic organs and reticuloendothelial system in aleukia, in: Alimentary Toxic Aleukia. Acta Chkalov Inst. Epidemiol. Microbiol. 2:125-144. First Communic. Chkalov.

4. Z. Alisova and S. Mironov (1944): On toxicity of prosomillet overwintered in the field, in: Data on Septic Angina. Proc. First Kharkov Med. Inst. a. Chkalov Inst. Epidemiol. Microbiol. 1:17-36. First Communic. Chkalov.

5. Z. I. Alisova (1947): General toxic action of cereal crops overwintered in the field, in: Alimentary Toxic Aleukia. Acta Chkalov Inst. Epidemiol. Microbiol. 2:104-118. First Communic. Chkalov.

6. Z. I. Alisova and S. G. Mironov (1947): General toxic action of cereal crops overwintered in the field, in: Alimentary Toxic Aleukia. Acta Chkalov Inst. Epidemiol. Microbiol. 2:97-103. Second Communic. Chkalov.

7. Z. I. Alisova (1947): General action of overwintered cereal extracts and toxic fungi on laboratory animals, in: Alimentary Toxic Aleukia. Acta Chkalov Inst. Epidemiol. Microbiol. 2:192 (Abstr.). Chkalov.

8. K. P. Andreyev (1939): Prosomillet poisoning in pastures. Soviet Vet. 9:75.

9. N. A. Antonov, G. S. Belkin, A. Z. Joffe, A. Ya. Lukin, and E. N. Simonov (1951): Feeding experiments on horses with cultures of the toxic fungi Fusarium poae (Peck.) Wr. and Cladosporium epiphyllum (Pers.). Acta Chkalov Agric. Inst. 4:47-56. Chkalov.

10. J. R. Bamburg, N. V. Riggs, and F. M. Strong (1968): The structures of toxins from two strains of Fusarium tricinctum. Tetrahedron 24:3329-3336.

11. J. R. Bamburg, W. F. O. Marasas, N. V. Riggs, E. B. Smalley, and F. M. Strong (1968): Toxic spiro-epoxy compounds from Fusarium and other Hyphomycetes. Biotechnol. Bioeng. 10:445-455.

12. J. R. Bamburg, F. M. Strong, and E. B. Smalley (1969): Toxins from moldy cereals J. Agric. Food Chem. 17:443-450.

13. J. R. Bamburg and F. M. Strong (1969): Mycotoxins of the Trichothecene family produced by Fusarium tricinctum and Trichoderma lignorum. Phytochemistry 8:2405-2410.

14. J. R. Bamburg and F. M. Strong (1971): In S. Kadis, A. Ciegler, and S. J. Ajl (eds.): Microbial toxins, Vol. 7: Algal and Fungal Toxins. Academic Press, New York, pp. 207-292.

15. G. L. Barer (1947): The problem of the chemical nature of the toxic cereals. Lecture on Republic Conference about Alimentary Toxic Aleukia, Moscow.

16. G. N. Beletskij (1945): Measures to prevent and fight against "septic angina" (alimentary toxic aleukia). Hyg. a. Sanit., Moscow, pp. 22-26.

17. V. I. Bilai (1947): Fusarium species on cereal crops and their toxic properties. Microbiology 16:11-17.

18. V. I. Bilai (1948): The action of extracts from toxic fungi on animal and plant tissues. Microbiology 17:142-147.

19. V. I. Bilai (1953): Toxic Fungi on the Grain of Cereal Crops. Publ. Acad. Sci. Ukr. S.S.R. Kiev, pp. 1-93.

20. V. I. Bilai (1955): Fusarium. Publ. Acad. Sci. Ukr. S.S.R. Kiev, pp. 1-319.

21. V. I. Bilai (1965): Toxins, in: Biologically Active Substances of Microscopic Fungi. Naukowa Dumka, Kiev, pp. 160-219.

22. H. R. Burmeister (1971): T-2 toxin production by Fusarium tricinctum on solid substrate. Appl. Microbiol. 21:739-742.

23. H. R. Burmeister and C. W. Hesseltine (1970): Biological assays for two mycotoxins produced by Fusarium tricinctum. Appl. Microbiol. 20:437-440.

24. H. R. Burmeister, J. J. Ellis, and C. W. Hesseltine (1972): Survey for Fusarium that elaborate T-2 toxin. Appl. Microbiol. 23:1165-1166.

25. E. A. Chernikov (1944): Problems of etiopathogenesis and therapy of "septic angina," in: Data of Septic Angina. Proc. First Kharkov Med. Inst. a. Chkalov Inst. Epidemiol. Microbiol. 1:173-183.

26. V. I. Chilikin (1944): Principal Problems of Clinical Syndromes, Pathogenesis, and Therapy of Alimentary Toxic Aleukia (Septic Angina). Reginal Publ. House, Kujbyshev, pp. 1-31.

27. V. I. Chilikin (1945): In A. I. Nesterov, A. H. Sysin, and L. N. Karlic
     (eds.): Clinical Aspects and Therapy of Alimentary Toxic Aleukia (Septic
     Angina). Publ. State Med. Lit. "MEDGIZ," pp. 39-54.
28. V. I. Chilikin (1947): Peculiarities and problems concerning clinical
     aspects, pathogenesis and therapy of alimentary toxic aleukia, in:
     Alimentary Toxic Aleukia. Acta Chkalov Inst. Epidemiol. Microbiol.
     2:145-151. Chkalov.
29. V. L. Davydova (1947): Sensibility of human and animal skin to toxin of
     cereals overwintered in the field, in: Alimentary Toxic Aleukia. Acta
     Chkalov Inst. Epidemiol. Microbiol. 2:94-96.
30. E. B. Davydovskij and A. G. Kestner (1935): On so-called septic angina
     (morphology and pathogenesis). Arch. Pathol. Anat. 1:11-30.
31. B. S. Drabkin and A. Z. Joffe (1950): The effect of extract from over-
     wintered cereals on Paramaecium caudatum. Acta Chkalov Med. Inst.
     2:92.
32. B. S. Drabkin and A. Z. Joffe (1952): The protistocide effect of certain
     mold. Microbiol. Acad. Sci. U.S.S.R. Moscow 21:700-704.
33. O. K. Elpidina (1945): The etiology of septic angina and determination
     of grain toxicity, in Z. I. Malkin (ed.): Data on Alimentary Toxic Aleukia.
     Kazan, pp. 78-82.
34. O. K. Elpidina (1945): Phytotoxin in grain causing septic angina. Lect.
     Acad. Sci. U.S.S.R. (DAN) 46:2.
35. O. K. Elpidina (1946): Biological methods of determining grain toxicity
     causing septic angina. Lect. Acad. Sci. U.S.S.R. 51:2.
36. O. K. Elpidina (1956): Biological and antibiotic properties of poin.
     Abstracts of Mycotoxicoses of Men and Agricultural Animals. Publ.
     Acad. Sci. Ukr. S.S.R. Kiev, pp. 23-24.
37. O. K. Elpidina (1959): Antibiotic and antiblastic properties of poin.
     Antibiotics 4:46.
38. O. K. Elpidina (1960): Toxic and antibiotic properties of poin, in V. I.
     Bilai (ed.): Mycotoxicoses of Man and Agricultural Animals. Publ.
     Acad. Sci. Ukr. S.S.R. Kiev, p. 59.
39. O. K. Elpidina (1961): The action of poin toxin from Fusarium sporo-
     trichioides v. poae on malignant tumor growth in experiments. Abstr.
     Report of the Conference on Mycotoxicoses, Kiev.
40. O. K. Elpidina (1958): The antiblastic properties of poin, according to
     experimental data. Kazan Med. J. 39:96-104.
41. J. Forgacs and W. T. Carll (1962): Mycotoxicoses. Adv. Vet. Sci. 7:
     273-382.
42. M. Yu. Friedman (1945): Prophylaxis of alimentary toxic aleukia (septic
     angina), in: Alimentary Toxic Aleukia (Septic Angina). Med. Publ. House,
     Moscow, p. 54.
43. V. G. Geimberg and A. M. Babusenko (1949): Dynamics of development
     of microflora in overwintered grain in experimental field conditions.
     Hyg. a. Sanit. 5:31-34.
44. N. B. Geminov (1945): Etiology of alimentary toxic aleukia, in V. Chilikin
     and N. Geminov (eds.): Septic Angina and its Treatment. Kuybyshev, pp.
     3-7.

45. N. B. Geminov (1945): Epidemiology of septic angina in the Kuybyshev district in 1945. Lecture on Republic Conference of Alimentary Toxic Aleukia, Moscow, December.

46. A. Genkin (1944): Condition of the upper respiratory tract in Septic Angina, in: Data on Septic Angina. Proc. First Kharkov Med. Inst. a. Chkalov Inst. Epidemiol. Microbiol. 1:117-128.

47. A. I. Germanov (1945): Some laboratory data relating to alimentary toxic aleukia. Sov. Med. 3:9.

48. W. Gerlach (1970): Suggestion for an acceptable modern Fusarium system, in E. A. Jamelainen (ed.): Ann. Acad. Sci. Fenn. Ser. A, IV Biologica, Helsinki 168:37-49.

49. G. Getsova (1960): Data on experimental study of fusariotoxicosis, in V. Bilai (ed.): Mycotoxicoses of Man and Agricultural Animals. Publ. Acad. Sci. Ikr. S.S.R. Kiev, p. 84.

50. M. W. Gilgan, E. B. Smalley, and F. M. Strong (1966): Isolation and partial characterization of a toxin from Fusarium tricinctum on moldy corn. Arch. Biochem. Biophys. 114:1-3.

51. W. O. Goldfredsen, J. F. Grove, and C. Tamm (1967): On the nomenclature of a new class of sesquiterpenoids. Helv. Chim. Acta 40:1666-1668.

52. W. L. Gordon (1952): The occurrence of Fusarium species in Canada. II. Prevalence and taxonomy of Fusarium species in cereal seed. Can. J. Bot. 30:209-251.

53. W. L. Gordon (1954): Taxonomy of Fusarium species in the seed of vegetables, forage, and miscellaneous crops. Can. J. Bot. 32:576-590.

54. W. L. Gordon (1960): The taxonomy and habitus of Fusarium species from tropical and temperate regions. Can. J. Bot. 38:643-658.

55. R. B. Gorodijskaja (1945): Blood changes induced by septic angina, in: Acta on Septic Angina. Bashkin Publ. House, Ufa, p. 112.

56. L. V. Gromashevskij (1945): Pathogenesis of alimentary toxic aleukia. J. Microbiol. Epidemiol. Immunol. 3:65.

57. M. D. Grove, S. G. Yates, W. H. Tallent, J. J. Ellis, I. A. Wolf, N. R. Kosuri, and R. E. Nicholas (1970): Mycotoxins produced by Fusarium tricinctum as possible causes of cattle disease. J. Agric. Food Chem. 18:734-736.

58. G. I. Grinberg (1943): Clinical aspects and pathogenesis of the so-called septic angina. Sov. Med. 10:24-34.

59. E. M. Gubarev and N. A. Gubareva (1945): Chemical nature and certain chemical properties of the toxic substances responsible for septic angina. Biochemistry 10:199-204.

60. Z. A. Gurewitch (1944): Neuropathologic peculiarities of septic angina, in: Data of Septic Angina. Proc. First Kharkov Med. Inst. a. Chkalov Inst. Epidemiol. Microbiol. 1:103-116.

61. E. T. Hrootski and A. Z. Joffe (1953): The action of Fusarium poae on the motor activity of the stomach of dogs. Acta Chkalov Agric. Inst. 6:59-62.

62. I.-C. Hsu, E. B. Smalley, F. M. Strong, and W. E. Ribelin (1972): Identification of T-2 toxin in moldy corn associated with a lethal toxicosis in cattle. Appl. Microbiol. 24:684-690.

63. A. Z. Joffe (1947): The biological properties of fungi, isolated from overwintered cereal crops, in: Alimentary Toxic Aleukia. Acta Chkalov Inst. Epidemiol. Microbiol. 2:192 (Abstr.). Chkalov.

64. A. Z. Joffe (1950): Toxicity of fungi on cereals overwintered in the field (on the etiology of alimentary toxic aleukia). Dissertation, Inst. Bot. Acad. Sci. U.S.S.R., Leningrad, p. 205.

65. A. Z. Joffe (1956): The etiology of alimentary toxic aleukia, in: Conference on Mycotoxicosis of Man and Husbandry Animals. Publ. Sci. Ukr. S.S.R. Kiev, pp. 36-38.

66. A. Z. Joffe (1956): The influence of overwintering on the antibiotic activity of several molds of the genus Cladosporium, Alternaria, Fusarium, Mucor, Thamnidium and Aspergillus. Acta Acad. Sci. Lithuanian S.S.R., Ser. B 3:85-96.

67. A. Z. Joffe (1956): The effect of environmental conditions on the antibiotic activity of some fungi of the genus Penicillium. Acta Acad. Sci. Lithuanian S.S.R., Ser. B. 4:101-114.

68. A. Z. Joffe (1960): Toxicity and antibiotic properties of some Fusarium. Bull. Res. Council Israel 8D:81-95.

69. A. Z. Joffe (1960): The mycoflora of overwintered cereals and its toxicity. Bull. Res. Council Israel 9D:101-126.

70. A. Z. Joffe (1962): Biological properties of some toxic fungi isolated from overwintered cereals. Mycopathol. Mycol. Appl. 16:201-226.

71. A. Z. Joffe (1963): Toxicity of overwintered cereals. Plant Soil 18:31-41.

72. A. Z. Joffe (1965): Toxin production by cereal fungi causing toxic alimentary aleukia in man, in G. N. Wogan (ed.): Mycotoxins in Foodstuffs. M.I.T. Press, Cambridge, Massachusetts, pp. 77-85.

73. A. Z. Joffe (1971): Alimentary toxic aleukia, in: Microbial Toxins, Vol. 7: Algal and Fungal Toxins. Academic Press, New York, pp. 139-189.

74. A. Z. Joffe (1973): Fusarium species of the Sporotrichiella section and relations between their toxicity to plants and animals. Pflkrankh. 80:92-99.

75. A. Z. Joffe (1974): Toxicity of Fusarium poae and F. sporotrichioides and its relation to alimentary toxic aleukia, in I. F. H. Purchase (ed.): Mycotoxins. Elsevier, Amsterdam, pp. 229-262.

76. A. Z. Joffe (1974): A modern system of Fusarium taxonomy. Mycopathol. Mycol. Appl. 53:201-230.

77. A. Z. Joffe (1974): Growth and toxigenicity of Fusarium of the Sporotrichiella section as related to environmental factors and culture substrates. Mycopathol. Mycol. Appl. 54:35-46.

78. A. Z. Joffe and S. G. Mironov (1947): Mycoflora of normal cereals and cereals overwintered in the field, Part 2, in: Alimentary Toxic Aleukia. Acta Chkalov Inst. Epidemiol. Microbiol. 2:35-41. Chkalov.

79. A. Z. Joffe and J. Palti (1974): Relations between harmful effects on plants and on animals of toxins produced by species of Fusarium. Mycopathol. Mycol. Appl. 52:209-218.

80. A. Z. Joffe and J. Palti (1975): Taxonomic study of Fusaria Sporotrichi-
    ella section used in recent toxicological work. Appl. Microbiol. 29:575-
    579.
81. V. M. Karatygin and Z. I. Rozhnova (1947): Vitamin insufficiency in
    alimentary toxic aleukia (septic angina). Sov. Med. 5:17.
82. L. N. Karlik (1945): The history of alimentary toxic aleukia (septic
    angina), in A. I. Nesterov, A. N. Sysin, and L. I. Karlic (eds.):
    Alimentary Toxic Aleukia (Septic Angina). Publ. State Med. Lit.,
    MEDGIZ, pp. 3-7.
83. N. I. Kolosova (1949): Data on chemical and toxic properties of some
    fatty acids isolated from grains causing alimentary toxic aleukia,
    Dissertation, Moscow, p. 112.
84. I. A. Kasirskij and G. A. Alekseyev (1948): Alimentary toxic aleukia,
    in: Disease of Blood and Hematopoietic System. MEDGIZ, pp. 204-217.
85. E. A. Kost (1935): Agranulocytosis and hemorrhagic angina. Sov. Clin.
    1:5.
86. E. N. Kovalev (1944): The nervous system in so-called septic angina.
    Neuropathol. Psychiatry 13:75.
87. M. A. Koza, I. A. Leontiev, and P. Ya. Yasnitskij (1944): Alimentary
    Toxic Aleukia, "Septic Angina." Med. Publ. House, Moscow, pp. 3-43.
88. N. I. Kozin and O. A. Yershova (1945): Cultivation method for deter-
    mining toxicity of cereal grain overwintered under snow (prosomillet).
    Proc. of Nutrition Inst. Acad. Med. Sci. U.S.S.R. Moscow, pp. 39-48.
89. V. L. Kretovich (1945): Biochemistry of grain causing septic angina.
    Abstr. Lect. on Republic Conference on Alimentary Toxic Aleukia,
    Moscow, pp. 9-10.
90. V. L. Kretovich and A. A. Bundel (1945): Investigations of the oil of
    toxic overwintered prosomillet. Biochemistry 10:216-224.
91. V. L. Kretovich, E. N. Mishustin, and A. A. Bundel (1946): Suppression
    of alcohol fermentation by products of oil decomposition. Biochemistry
    11:149-154.
92. V. L. Kretovich and N. I. Sosedov (1945): Biochemical properties of
    toxic prosomillet. Biochemistry 10:279-284.
93. V. L. Kretovich and Z. G. Skripkina (1945): Diagnostics of overwintered
    toxic grains in the field. Lect. Acad. Sci. U.S.S.R. 47:504-407.
94. V. T. Kudryakov (1945): Pathogenesis and treatment of alimentary toxic
    aleukia. Clin. Med. 24:14.
95. E. S. Kvashnina (1948): Mycoflora of cereal crops overwintered in the
    field, in: Data on Cereal Crops Wintered under Snow. Publ. Minist.
    Agr. U.S.S.R. Moscow, pp. 86-92.
96. Ya. Kh. Lando (1935): On the pathologic anatomy of septic angina. Med.
    J. Kazakhstan 4-5:88-91.
97. Ya. Kh. Lando (1939): Material relating to the pathologic anatomy of
    septic angina. Sov. Publ. Health Serv. Kirghiz S.S.R. 6:72-86.
98. I. I. Levin (1946): Treatment of alimentary toxic aleukia with sulfamid
    compound preparations. Clin. Med. 7:54.

99. L. A. Lindenfelser, E. B. Lillehoj, and H. R. Burmeister (1974): Afla-
     toxin and trichothecene toxins: Skin tumor induction synergistic acute
     toxicity in white mice. J. Natl. Cancer Inst. 52:113-116.

100. D. S. Lovla (1944): Septic angina in the Chkalov district, in: Alimentary
     Toxic Aleukia (Septic Angina). Abstr. Lect. Publ. Lab. Septic Angina,
     Chkalov Inst. Epidemiol. Microbiol., Kharkov Med. Inst. a. Clinic.
     Hosp. of Orenburg Railway. Chkalov, p. 1.

101. N. N. Lozanov and V. Yà. Tsareva (1944): Septic Angina. Tatar
     A.S.S.R. Publ. House, Kazan, pp. 3-15.

102. A. Ya. Lukin and M. G. Berlin (1947): Toxic influences of overwintered
     prosomillet on the organs of horses and pigs. Proc. Chkalov Agric.
     Inst. 3:65-71.

103. A. Ya. Lukin, N. A. Antonov, and I. N. Simonov (1947): Feeding tests
     with toxic cereals on pigs and sheep. Proc. Chkalov Agric. Inst. 3:78-86.

104. A. Ya. Lukin, N. A. Antonov, and I. N. Simonov (1947): Feeding tests
     with toxic cereals on horses. Proc. Chkalov Agric. Inst. 3:93-106.

105. M. A. Lyass (1940): Agranulocytosis. Vitebsk Med. Inst., pp. 3-95.

106. G. I. Maisuradge (1953): The role of fungi in developing toxicosis in
     horses by feeding them germinating oats. Abstr. Dissert. Moscow,
     pp. 1-15.

107. Z. I. Malkin and N. N. Odelevskaja (1945): Clinical aspects and treat-
     ment of alimentary toxic aleukia (septic angina), in: Alimentary Toxic
     Aleukia. Kazan, p. 19.

108. E. M. Manburg (1944): Clinical aspects of Septic Angina, in: Data on
     Septic Angina. Proc. First Kharkov Med. Inst. a. Chkalov Inst. Epi-
     demiol. Microbiol. 1:85-102. Chkalov.

109. E. M. Manburg and E. A. Rachalskij (1947): Clinical aspects and
     therapy of alimentary toxic aleukia, in: Alimentary Toxic Aleukia. Acta
     Chkalov Inst. Epidemiol. Microbiol. 2:152-163.

110. O. S. Manoilova (1947): Chemical diagnosis of toxic overwintered
     cereals. Lect. Republ. Conf. on Alimentary Toxic Aleukia, Moscow.

111. W. F. O. Marasas, J. R. Bamburg, E. B. Smalley, F. M. Strong,
     W. L. Ragland, and P. E. Degurse (1969): Toxic effects on trout, rats
     and mice of T-2 toxin produced by the fungus Fusarium tricinctum (Cd.)
     Snyd. and Hans. Toxicol. Appl. Pharmacol. 15:471-482.

112. W. F. O. Marasas, E. B. Smalley, P. E. Degurse, P. E. Bamburg,
     J. R. and R. E. Nichols (1967): Acute toxicity to rainbow trout (Salmo
     gairdnerii) of a metabolite produced by the fungus Fusarium tricinctum.
     Nature (London) 214:817-818.

113. T. Matuo (1972): Taxonomic studies of phytopathogenic Fusarium in
     Japan. Rev. Plant Protect. Res. (Tokyo) 5:34-45.

114. C. F. Mayer (1953): Endemic panmyelotoxicosis in the Russian grain
     belt. Part I. The clinical aspects of alimentary toxic aleukia (ATA).
     A comprehensive review. Military Surgeon 113:173-189.

115. C. F. Mayer (1953): Endemic panmyelotoxicosis in the Russian grain
     belt. Part II. The botany, phytopathology, and toxicology of Russian
     cereal food. Military Surgeon 113:295-315.

116. C. J. Mirocha and S. Pathre (1973): Identification of the toxic principle in a sample of poaefusarin. Appl. Microbiol. 26:719-724.

117. S. Mironov (1945): Etiology of septic angina (alimentary toxic aleukia) and measures for its prevention. J. Microbiol. Epidemiol. Immunol. 6:70-77.

118. S. Mironov and R. Fok (1944): Toxicity of overwintered prosomillet in the field. Sec. Commun., in: Data on Septic Angina. Proc. First Kharkov Med. Inst. a. Chkalov Inst. Epidemiol. Microbiol. 1:37-46.

119. S. Mironov, R. Soboleva, R. Fok, and Yudenich (1944): Phytopathological analysis of overwintered prosomillet samples in the field, in: Data on Septic Angina. Proc. First Kharkov Med. Inst. a. Chkalov Inst. Epidemiol. Microbiol. 1:47-52.

120. S. G. Mironov, A. Z. Joffe, M. K. Bakbardina, R. Fok, and V. L. Davydova (1947): Phytopathological analysis of toxic overwintered prosomillet samples. Second communic., in: Alimentary Toxic Aleukia. Acta Chkalov Inst. Epidemiol. Microbiol. 2:11-18.

121. S. G. Mironov and V. A. Myasnikov (1947): Characteristics of toxins from overwintered cereals in the field. First communic., in: Alimentary Toxic Aleukia. Acta Chkalov Inst. Epidemiol. Microbiol. 2: 61-65.

122. S. G. Mironov and Z. I. Alisova (1947): Detection of toxic compounds of fungal derivation in overwintered cereals by means of immunity tests, in: Alimentary Toxic Aleukia. Acta Chkalov Inst. Epidemiol. Microbiol. 2:192.

123. S. G. Mironov and V. L. Davydova (1947): Sensibility of man and animal skin to toxins of cereals overwintered in the field, in: Alimentary Toxic Aleukia. Acta Chkalov Inst. Epidemiol. Microbiol. 2:89-93.

124. S. G. Mironov and R. A. Fok (1947): Skin tests of rabbits for determination of toxicity of overwintered cereals, in: Alimentary Toxic Aleukia. Acta Chkalov Inst. Epidemiol. Microbiol. 2:80-82. Second communic. Chkalov.

125. S. G. Mironov and R. A. Fok (1947): Skin test of rabbits for determination of toxicity of overwintered cereals, in: Alimentary Toxic Aleukia. Acta Chkalov Inst. Epidemiol. Microbiol. 2:83-88. Third communic. Chkalov.

126. S. G. Mironov, A. I. Strukov, and R. A. Fok (1947): Skin test of rabbits for determination of toxicity of overwintered cereals, in: Alimentary Toxic Aleukia. Acta Chkalov Inst. Epidemiol. Microbiol. 2:73-79. First communic. Chkalov.

127. S. G. Mironov and A. Z. Joffe (1947): The dynamics of toxin accumulation in overwintered cereals in the field, in: Alimentary Toxic Aleukia. Acta Chkalov Inst. Epidemiol. Microbiol. 2:19-22. First communic. Chkalov.

128. S. G. Mironov and A. Z. Joffe (1947): The dynamics of toxin accumulation in overwintered cereals in the field, in: Alimentary Toxic Aleukia. Acta Chkalov Inst. Epidemiol. Microbiol. 2:23-27. Second communic. Chkalov.

129. E. N. Mishstin, V. L. Kretovich, and A. A. Bundel (1946): The fermentation test as a diagnostic method for toxic overwintered prosomillet. Hyg. Sanit. 11:32-35.

130. K. E. Murashkinskij (1934): On the study of fusariosis of cereal crops. Species of genus Fusarium on cereal crops in Siberia. Proc. Siber. Agric. Acad. 3:87-114. Osmk.

131. A. L. Myasnikov (1935): Clinical aspects of alimentary hemorrhagic aleukia, in: Alimentary Hemorrhagic Aleukia (Septic Angina). West. Siber. Territor. Publ. Health Serv., Novosibirst, p. 48.

132. V. A. Myasnikov (1948): Characteristics of extracts of various fungi treated by the barium method, compared to extracts of prosomillet exposed in winter. Chkalov Inst. Epidemiol. Microbiol. (MS).

133. V. A. Myasnikov (1947): Experiments in destruction and neutralization of toxic material from overwintered prosomillet, in: Alimentary Toxic Aleukia. Acta Chkalov Inst. Epidemiol. Microbiol. 2:55-57. Chkalov.

134. M. I. Nakhapetov (1944): Septic Angina. Leningrad, MEDGIZ, p. 14.

135. V. S. Nesterov (1948): The clinical aspects of septic angina. Clin. Med. 7:34.

136. N. V. Okuniev (1945): New data on the chemical nature of the initial toxicity of cereal grain causing alimentary toxic aleukia. Abstr. Republic Conference on Alimentary Toxic Aleukia, Moscow, p. 15.

137. A. P. Okuniev and V. A. Naumov (1943): Toxicity of overwintered cereals. Proc. Kirov Zootechnol. Vet. Inst. 5:110-119.

138. L. E. Olifson (1956): The chemical activity of Fusarium sporotrichiella, Abstr. on Mycotoxicoses in Human and Agricultural Animals. Publ. Acad. Sci. Ukr. S.S.R. Kiev, pp. 22-22.

139. L. E. Olifson (1957): Toxins isolated from overwintered cereals and their chemical nature. Monitor, Orenburg Sect. of the U.S.S.R. D. J. Mendelijev Chem. Soc. 7:21-35.

140. L. E. Olifson (1957): Chemical action of some fungi on overwintered cereals. Monitor, Orenburg Sect. of the U.S.S.R. D. J. Mendeleyev Chem. Soc. 7:37-45.

141. L. E. Olifson (1955): New chemical methods of determining toxicity of cereal crops. Communic. on Scient. Works of members D.J. Mendeleyev All Soviet Chem. Soc. Publ. Acad. Sci. U.S.S.R. Moscow, pp. 58-59.

142. L. E. Olifson (1965): Chemical and biological properties of toxic materials derived from grain infected with the fungus Fusarium sporotrichiella. Moscow Technol. Inst. Industr. Nutrition. Abstr. Dissertation, p. 36.

143. L. E. Olifson (1972): The problems of toxic steroids in microscopic fungi, Fusarium sporotrichiella. Proc. Sympos. Mycotoxins. Acad. Sci. Ukr. S.S.R. Kiev, p. 12.

144. L. E. Olifson, B. S. Drabkin, and A. Z. Joffe (1950): The influence of Fusarium fungi on millet oil. Acta Chkalov Med. Inst. 2:103-107.

145. L. E. Olifson and A. Z. Joffe (1954): On the changes in some chemical constants of millet oil under the influence of fungi developing on millet. Monitor, Orenburg Sect. of the U.S.S.R. D. J. Mendeleyev Chem. Soc. 5:61-65.

146. L. E. Olifson, S. M. Kenina, and V. L. Kartashova (1972): Chromatographic method to identify toxicity of grain of cereals (wheat, rye, millet
and others) affected by a toxigenic strain of Fusarium sporotrichiella.
Instruction on how to identify the toxicity of the grain. Monitor, Orenburg
Sec. of the U.S.S.R. D. J. Mendeleyev Chem. Soc., pp. 3-8.

147. I. S. Pentman (1935): Pathologic Anatomy of Alimentary Hemorrhagic
Aleukia (Septic Angina). West-Siber. Region. Health, Novosibirsk, pp.
17-29.

148. G. M. Peregud (1947): Clinical aspects and treatment of oral cavity and
upper respiratory tract in cases of alimentary toxic aleukia, in Alimentary Toxic Aleukia. Acta Chkalov Inst. Epidemiol. Microbiol. 2:170-175.

149. M. M. Pidoplichka and V. I. Bilai (1946): Toxic fungi on cereal grains.
Publ. Acad. Sci. Ukr. S.S.R. Kiev, pp. 65.

150. A. S. Poznanski (1947): Neuropsychic disturbances in alimentary toxic
aleukia, in: Alimentary Toxic Aleukia. Acta Chkalov Inst. Epidemiol.
Microbiol. 2:176-178. Chkalov.

151. A. I. Raillo (1950): Fungi of the genus Fusarium. Publ. State Agric. Lit.
Moscow, p. 415.

152. A. V. Reisler (1943): Septic Angina. Publ. Med. House, Moscow, p. 19.

153. A. V. Reisler (1952): Alimentary toxic aleukia, in: Hygiene of Nutrition.
Med. State Publ. House, pp. 402-405.

154. E. D. Romanova (1947): Clinical observation and therapy of alimentary
toxic aleukia, in: Alimentary Toxic Aleukia. Acta Chkalov. Inst.
Epidemiol. Microbiol. 2:164-169. Chkalov.

155. Yu. I. Rubinstein (1948): Microflora of cereals overwintered under snow.
Proc. Nutrition Inst. Acad. Med. Sci. U.S.S.R. Moscow, pp. 29-38.

156. Yu. I. Rubinstein (1950): Biochemical properties of Fusarium cultures
(sect. Sporotrichiella). Microbiology 19:438-443.

157. Yu. I. Rubinstein (1951): Some properties of the Fusarium sporotrichioides toxin. Acta Acad. Med. Sci. U.S.S.R. Nutr. Probl. 13:247-253.

158. Yu. I. Rubinstein (1953): The etiology of Urovsk disease. Nutr. Probl.
12:73-81.

159. Yu. Rubinstein (1956): Actual problems in study of fusariotoxicoses.
Nutr. Probl. 15:8.

160. Yu. I. Rubinstein (1960): Fusariotoxicoses in food, in V. I. Bilai (ed.):
Mycotoxicoses of Man and Agricultural Animals. Publ. Acad. Sci. Ukr.
S.S.R. Kiev, p. 71.

161. Yu. I. Rubinstein, Yu. Kykel, and G. Kudinova (1961): New experimental
chronic toxicosis caused by Fusarium sporotrichiella. Abstr. Lect. II
Conference on Mycotoxicoses, Kiev, p. 16.

162. Yu. I. Rubinstein and L. S. Lyass (1948): On the etiology of alimentary
toxic aleukia (septic angina). Hyg. Sanit. 7:33-38.

163. Riazanov (1947): Lecture on the Republic Conference on Alimentary Toxic
Aleukia, Moscow.

164. Riazanov (1948): Lecture on the Republic Conference on Alimentary Toxic
Aleukia, Moscow.

165. M. Saito and T. Tatsuno (1971): Toxins of Fusarium nivale, in S. Kadis,
      A. Ciegler, and S. J. Ajl (eds.): Microbial Toxins, Vol. 7: Algal and
      Fungal Toxins. Academic Press, New York, pp. 293-316.

166. A. Kh. Sarkisov (1944): Method of determining toxicity of cereal crops
      overwintered in the field. Hyg. Sanit. 9:19-22.

167. A. Kh. Sarkisov (1948): Data on toxicity of cereals overwintered under
      snow, in: Cereal Crops Overwintered under Snow. Publ. Minist. Agric.
      U.S.S.R., Moscow, pp. 22-40.

168. A. Kh. Sarkisov (1950): Etiology of "septic angina" in man. J. Microbiol.
      Epidemiol. Immunol. 1:43-47.

169. A. Kh. Sarkisov (1954): Mycotoxicoses. Publ. Minist. Agric. U.S.S.R.,
      Moscow, p. 216.

170. A. Kh. Sarkisov, N. E. Korneyev, E. S. Kvashnina, V. P. Koroleva,
      P. A. Gerasimova, and N. S. Akulova (1948): The harm caused by over-
      wintered cereal crops to livestock and poultry, in: Cereal Crops Wintered
      under Snow. Publ. Minist. Agric. U.S.S.R., Moscow, pp. 10-21.

171. A. Kh. Sarkisov and E. S. Kvashnina (1948): Toxico-biological properties
      of Fusarium sporotrichioides, in: Cereal Crops Wintered under Snow.
      Publ. Minist. Agric. U.S.S.R., Moscow, pp. 86-92.

172. R. Schoental and A. Z. Joffe (1974): Lesions induced in rodents by extracts
      from cultures of Fusarium poae and F. sporotrichioides. J. Pathol.
      112:37-42.

173. P. M. Scott and E. Somers (1969): Biologically active compounds from
      field fungi. J. Agric. Food Chem. 17:430-436.

174. E. Seemüller (1968): Untersuchungen über die morphologische und
      biologische Differenzierung in der Fusarium, Section Sporotrichiella.
      Mitt. Biol. Bundesanst., Berlin-Dahlem 127:1-93.

175. P. G. Sergiev (1945): Epidemiology of alimentary toxic aleukia, in A. I.
      Nesterov, A. N. Sysin, and L. I. Karlic (eds.): Alimentary Toxic Aleukia
      (Septic Angina). Publ. State Med. Lit. MEDGIZ, pp. 7-11.

176. P. G. Sergiev (1946): Harvesting losses and overwintered grain in the
      field — source of disease. Kolkhoz. Product 7:16.

177. P. G. Sergiev (1948): Septic angina, in: Artic. Lect. Conversat. on Sanit.,
      Instructive Subjects. Barnaul, pp. 62-66.

178. B. M. Serafimov (1945): Mental disorder, caused by alimentary toxic
      aleukia. Neuropathol. Psychiatry, Moscow 5:25-30.

179. B. N. Serafimov (1946): Mental symptomatology of alimentary toxic aleukia.
      Neuropathol. Psychiatry 6:50-53.

180. R. S. Shklovskaja and F. P. Brodskaja (1944): The so-called "septic angina,"
      in G. S. Lurie (ed.): Date Scient.-Pract. Works of Town Sterlitamak
      Physicians. Sterlitamak, Bashkir S.S.R., pp. 5-31.

181. O. N. Sirotinina (1945): Toxicity of cereals overwintered in the field.
      Dissertation Saratov, p. 232.

182. S. I. Slonevskij (1946): Alimentary toxic aleukia (septic angina) in Udmurt
      A.S.S.R. Hyg. Sanit. 6:23.

183. E. B. Smalley, W. F. O. Marasas, F. M. Strong, J. R. Bamburg, R. E.
Nichols, and N. R. Kosuri (1970): Mycotoxicoses associated with moldy
corn, in M. Herzberg (ed.): Toxic Microorganisms. U.S. Dept. of the
Interior, Washington, D.C., pp. 163-173.

184. V. A. Smirnova (1945): Long-term consequences of septic angina in laryngeal organs, in: Data on Septic Angina. Ufa, p. 145.

185. W. C. Snyder and H. N. Hansen (1945): The species concept in Fusarium
with reference to Discolor and other sections. Am. J. Bot. 32:657-666.

186. A. Strukov and M. Tishchenko (1944): Pathomorphology of septic angina,
in: Data on Septic Angina. Proc. First Kharkov Med. Inst. a. Chkalov
Inst. Epidemiol. Microbiol. 1:53-84. Chkalov.

187. A. I. Strukov (1947): Pathological changes in animal tissues caused by
toxin from overwintered cereals, in Alimentary Toxic Aleukia. Acta
Chkalov Inst. Epidemiol. Microbiol. 2:117-119. Sec. communic. Chkalov.

188. A. I. Strukov and S. G. Mironov (1947): Pathological changes in animal
tissues caused by toxin from overwintered cereals, in: Alimentary Toxic
Aleukia. Acta Chkalov Inst. Epidemiol. Microbiol. 2:109-116. First
communic. Chkalov.

189. A. I. Strukov and M. A. Tishchenko (1947): Some suppositions on patho-
morphology and pathogenesis of alimentary toxic aleukia, in: Alimentary
Toxic Aleukia. Acta Chkalov Inst. Epidemiol. Microbiol. 2:120-124.
Chkalov.

190. E. D. Svojskaja (1947): Experiments in isolating toxic substances from
overwintered prosomillet, in: Alimentary Toxic Aleukia. Acta Chkalov
Inst. Epidemiol. Microbiol. 2:45-54. Chkalov.

191. H. Stähelin, M. E. Kalberer, E. Signer, and S. Lazáry (1968): Uber
einige biologische Wirkungen des Cytostaticum Diacetoxyscirpenol.
Arzneim. Forsch. 18:989-994.

192. B. T. Talayev, B. I. Mogunov, and E. N. Sharbe (1936): Alimentary
hemorrhagic aleukia. Nutr. Probl. 5-6:27-35.

193. D. I. Tatarinov (1945): Problems to be discussed in the clinical aspects
and therapy of alimentary toxic aleukia, in: Acta of Septic Angina. Ufa,
p. 12.

194. G. N. Teregulov (1945): Clinical aspects and treatment of alimentary toxic
aleukia by propaeduetic-therapy materials of Bashkir Med. Inst., in: Data
on Septic Angina. Ufa, p. 30.

195. T. Ya. Tkachev (1945): The fight against septic angina. Hyg. Sanit. 6:24.

196. M. V. Tomina (1948): Distribution of fungal toxin from overwintered
cereals in animal organs. Abstr. Republic Conference on Alimentary Toxic
Aleukia, Moscow, p. 10.

197. H. Tookey, S. G. Yates, J. J. Ellis, M. D. Grove, and R. E. Nochols
(1972): Toxic effects of a butenolide mycotoxin and of Fusarium tricinctum
cultures in cattle. J. Am. Vet. Med. Assoc. 60:1522-1526.

198. Y. Ueno, N. Sato, K. Iishii, K. Sakai, and M. Enomoto (1972): Toxico-
logical approaches to the metabolites of Fusaria. V. Neosolaniol;
T-2 toxin and butenolide, toxic metabolites of Fusarium sporotrichioides
NRRL 3510 and Fusarium poae 3287. Jap. J. Exp. Med. 42:461-472.

199. Y. Ueno, N. Sato, K. Ishii, K. Sakai, H. Tsunoda, and M. Enomoto (1973): Biological and chemical detection of trichothecene mycotoxins of Fusarium species. Appl. Microbiol. 25:699-704.
200. G. M. Veindrach and S. V. Fadeyeva (1937): The blood-picture in septic granulocytic angina. Kazan Med. J. 9:1065-1072.
201. K. I. Vertinskij and V. A. Adutskevich (1948): Pathomorphologic studies of cats who died by experimental alimentary toxic aleukia, in Cereal Crops Wintered under Snow. Publ. Minist. Agric. U.S.S.R., pp. 80-86.
202. H. W. Wollenweber and O. A. Reinking (1935): Die Fusarien, ihre Beschreibung, Schadwirkung und Bekämpfung. P. Parey, Berlin, p. 355.
203. S. G. Yates, H. L. Tookey, J. J. Eliss, W. H. Tallent, and I. A. Wolff (1969): Mycotoxins as a possible cause of fescue toxicity. J. Agric. Food Chem. 17:437-442.
204. S. G. Yates, H. L. Tookey, and J. J. Ellis (1970): Survey of tall-fescue pasture: Correlation of toxicity of Fusarium isolates to known toxins. Appl. Microbiol. 19:103-105.
205. V. V. Yefremov (1944): On the so-called alimentary toxic aleukia ("septic angina"). Sov. Med. 1-2:19-21.
206. V. V. Yefremov (1944): Alimentary toxic aleukia (septic angina). Hyg. Sanit. 7-8:18-45.
207. V. V. Yefremov (1948): Alimentary Toxic Aleukia. Publ. Med. Lit. Moscow, p. 120.
208. V. Yudenich, C. Mironov, P. Soboleva, and R. Fok (1944): The septic angina in the Chkalov district in 1944, in: Data on Septic Angina. Proc. First Kharkov Med. Inst. a. Chkalov Epidemiol. Microbiol. 1:5-16. Chkalov.
209. A. P. Zavyalova (1946): Chemical and toxicological characteristics of lipoprotheids isolated from wheat, causing alimentary toxic aleukia. Dissertation, Kuybyshev, p. 164.
210. B. Ya. Zhodzishkij (1933): Data on the study of clinical aspects, pathogenesis and alimentary panhematopathy treatment in so-called septic angina (agranulocytosis, hemorrhagic aleukia, panmyelophthisis). Dissertation, Novosibirsk, p. 137.
211. V. A. Zhukhin (1945): Data on pathogenic anatomy and pathogenesis in so-called septic angina (alimentary toxic aleukia), in: Acta of Septic Angina. Ufa, p. 82.
212. B. Yagen and A. Z. Joffe (1976): Screening of toxic isolates of Fusarium poae and F. sporotrichioides involved in causing Alimentary Toxic Aleukia. Appl. Environ. Microbiol. 32:423-427.
213. A. Z. Joffe and B. Yagen (1977): Comparative study of the yield of T-2 toxin produced by Fusarium poae, F. sporotrichioides, and F. sporotrichioides var. tricinctum from different sources. Mycopathologia 60:1-5.

Abraham Z. Joffe

3.11   Human Stachybotryotoxicosis*

3.11.1   Introduction

Stachybotrys alternans Bon. [synonym S. atra Corda, S. chartarum (Ehren-
berg ex Link) Huges] (Sec. 1.6) is a common saprophyte, growing well on
substances rich in cellulose, such as straw or hay. The black color of this
fungus is evident in heavily contaminated material as a sooty layer on the
straw, arousing suspicion of its presence. About two-thirds of the isolates
have been proven capable of producing toxin under experimental conditions
[4, 8].

Evidence has been obtained showing that straw or hay contaminated by
Stachybotrys fungi may be dangerous to people handling the material.

3.11.2   Organ Systems Affected and
Symptomatology

In the reviews of Sarkisov et al. [6] and Forgacs [2] references are found
relating to the risks to those persons handling Stachybotrys-contaminated
material. These authors describe the following symptoms, among others:
cough, rhinitis, burning sensation in the mouth and nasal passages, and
cutaneous irritation at the point of toxin contact.

According to Gajdusek [3] cases of human stachybotryotoxicosis have
mainly been observed in the regions in which the equine disease has also been
reported. The people affected by stachybotryotoxicosis had handled S.
alternans-contaminated hay or straw in the feeding of animals. Gajdusek also
states that human stachybotryotoxicosis has been reported in regions in which
hay or straw has been used for bedding or as fuel to heat homes. In man, the
symptoms were first a rash at points with heavy perspiration, such as the
armpits. Moist dermatitis is the next stage of stachybotryotoxicosis, and this
is usually succeeded by a phase characterized by dried, crusted layers of a
serous exudate. "Catarrhal angina," with painful, severe pharyngitis and a
burning sensation in the nose follows. The nasal exudate may be bloody.
Fever may occur in rare instances, and the cough varies from moderate to
severe. Leukopenia has been observed in some patients.

Ožegović et al. [5], in their paper on stachybotryotoxicosis in young
cattle, state that farm workers who were handling Stachybotrys-contaminated

---

*See Sec. 3.7.6.1 for a discussion of Sporidesmins in man.

straw developed a disease within 2 to 3 days. There was strong itching and reddening of the skin, especially on the scrotum, and later around the anus and on the medial faces of the thighs. Within 3 days, vesicles of millet corn size, containing pus, developed in the skin. Both men and women displayed symptoms of rhinitis, pharyngitis, and conjunctivitis. Vesicles developed in the mucous membranes of the nose and later dark crusts. The laryngitis and pharyngitis were quite heavy in some of the workers, and one woman lost her voice completely. The disease had an average duration of 3 weeks. According to Ožegović and his co-workers the disease was caused by a direct effect of the toxic substance connected with the fungal spores.

Forgacs [2] has reviewed an article by Vertinskii, describing human stachybotryotoxicosis as a dermatitis localized preeminently on the scrotum and in the axillary regions, less frequently in other parts of the body.

According to Drobotko, as reviewed by Gajdusek [3], stachybotryotoxicosis in humans appears not only as a dermatitis, but occasionally also as a general toxicosis brought about by absorption of toxic substances through the skin or by inhalation of the mold dust.

Szabó and co-workers [7] in their report on stachybotryotoxicosis in pigs, also include information on human stachybotryotoxicosis. Six of the attendants who handled Stachybotrys-contaminated straw fell ill. The nose became swollen and scaly, and some of the attendants had nosebleeds. One of the workers also suffered from conjunctivitis.

Dzhilavyan [1] has described essentially the same symptoms as Gajdusek [3], in people handling Stachybotrys-contaminated straw. In addition to dermatitis, catarrhal angina, rhinitis, and conjunctivitis, the author mentions toxic lesions on the face, axillae, rear face of the knee joint, and perineum; women may also present an affection around the mammae. In addition to cough, rhinitis, and nosebleeds, Dzhilavyan also described moderate fever, headache, general feebleness, and fatigue as symptoms of human stachybotryotoxicosis.

3.11.3   Control

The removal of contaminated material from the environment results in a rapid recovery in mild affections. Symptomatic treatment may prove beneficial.

Persons who come into contact with Stachybotrys material in laboratories should take extremely good care of their protective apparel. The eyes should be protected by goggles, the respiratory organs by masks, and the hands by gloves. Toxin dissolved in organic solvents such as ether or acetone, should be handled with particularly great care. These solvents appear to promote the penetration of the toxins through the superficial layers of the body.

References

1. H. A. Dzhilavyan (1963): Stachybotryotoxicosis, in A. N. Bakulev (ed.):
    Great Medical Encyclopaedia, Vol. 31. Soviet Encyclopaedia, State
    Scientific Edition. Moscow, pp. 360-363 (in Russ.).

2. J. Forgacs (1972): Stachybotryotoxicosis, in S. Kadis, A. Ciegler, and S.
    J. Ajl (eds.): Microbial Toxins, 1st ed., Vol. 8. Academic Press, New
    York, pp. 95-128.

3. D. C. Gajdusek (1953): Stachybotryotoxicosis. Acute infectious hemor-
    rhagic fevers and mycotoxicoses in the Union of Soviet Socialistic
    Republics. Med. Sci. Publ. No 2. Army Med. Serv. Grad. School, Walter
    Reed Army Medical Center, Washington, D.C., pp. 107-111.

4. E.-L. Korpinen and J. Uoti (1974): Studies on Stachybotrys alternans II.
    Occurrence, morphology and toxigenicity. Acta Pathol. Microbiol. Scand.
    Sec. B. 82:1-6.

5. L. Ožegović, R. Pavlović, and B. Milošev (1971): Toxic dermatitis, con-
    junctivitis, rhinitis, pharyngitis and laryngitis in fattening cattle and farm
    workers caused by molds from contaminated straw (stachybotryotoxicosis?).
    Veterinaria (Sarajevo) 20:263-267 (Serbocroat).

6. A. H. Sarkisov, V. P. Koroleva, E. S. Kvashnina, and V. F. Grezin
    (1971): Diagnosis of the Fungal Diseases in Animals. Mycosis and Myco-
    toxicosis. Kolos, Moscow, pp. 84-91 (Russ.).

7. I. Szabó, F. Rátz, P. Áldásy, P. Szabó, and L. Gaál (1970): Scaly and
    scabious skin affection and rhinitis (stachybotryotoxicosis) in pig stocks.
    I. Clinical observations and aetiological investigations. Magy. Állatorv.
    Lapja 25:21-26 (Hung.).

8. R. V. Yuskiv (1969): Toxicity of different strains of Stachybotrys alternans
    Bonord. and Stachybotrys lobulata Berk. Mikrobiol. Zh. (Kiev) 31:27-31
    (U.).

E.-L. Hintikka
(née Korpinen)

3.12 The Role of Mycotoxins in Human Pulmonary Disease

3.12.1 Introduction

Ramazzini [1], the father of occupational medicine, accurately described diseases of workers inhaling "foul and mischievous powder" from handling food, fodder, and fiber crops in 1705. In recent decades such pulmonary diseases have been causally related to fungi, particularly thermophilic actino-myces growing on these natural products. To give some perspective to diseases which may be due to mycotoxins it appears useful to make a primary division of mycotic pulmonary diseases based on whether fungi grow in the host, the mycoses, or fungal products cause ill effects, the toxomycoses. Human mycoses include histoplasmosis, blastomycosis, coccidioidomycosis, and opportunistic fungal infections which constitute challenging and perplexing medical problems. The term "toxomycosis" has been applied to diseases produced by inhalation of fungal spores, mycelia, or the decaying matter upon which saprophytes (fungi) are growing. Effects are produced without growth of fungi in the host. Samsonov [2] classified diseases resulting from the absorption of fungal toxins through the mucous membranes of the respiratory tract as toxomycoses. Kováts and Bugyi [3, see also 4] extended the term to include the alveolar reactions which are called hypersensitivity pneumonitis or extrinsic allergic alveolitis [5]. The mechanism of these diseases appears to include the toxic effects of fungal products and host defensive, immune responses.

Fungal toxins may be responsible for much of the damage produced by either growing fungi or inhaled fungal products. Thus the mycotic and toxo-mycotic mechanisms may overlap in some instances as they do in disease manifestations due to certain Aspergillus species. Further complexity is contributed by host defenses, particularly those involving cellular memory and modification of response, the immune system. Not only may immune cascades perpetuate responses but these may vary, depending on the site in the lung and the activity of other defense systems. In brief, what follows concerning the effects of fungi on the respiratory tract must be regarded as a tentative synthesis of available data organized with the viewpoint that human hosts have had insufficient evolutionary preparation (immunity) to deal with large quantities of inhaled fungal toxins just as they usually lack well-modulated defenses to handle inhaled synthetic chemicals.

91

3.12.2   Types of Mycotic Pulmonary Disease

3.12.2.1   Mycoses

Infections may occur to either invasive or saprophytic fungi. The former
group includes diseases caused by the Deuteromycetes and by the Actinomycetaceae. Deuteromycetes, imperfect fungi, include Histoplasma,
Coccidioides, Blastomyces, and Cryptococcus which grow in tissue in the
yeast form. Usually, these fungi are primary pathogens producing localized
infections in the lung of a relatively benign nature. In association with
resistance-lowering diseases, e.g., lymphoma, leukemia, Hodgkin's disease,
cancer, diabetes mellitus, or during chemotherapy or immunosuppressive
therapy and occasionally unassociated with recognized disease, generalized
dissemination occurs. Epidemiological, immunological, and clinical aspects
have been well defined [6-9]. Although there are scattered clues as to the
mechanism of these diseases, their chemical composition, particularly their
mycotoxin content or elaboration, has received little study.
    The Actinomycetaceae include Actinomyces, Streptomyces, and Nocardia.
They grow as mycelia, usually in the mouth or gastrointestinal tract, and are
invasive only when there is a major break in the host defenses. Although
actinomycosis may be localized in the lung and produce diagnostic difficulties
in distinguishing it from bronchogenic carcinoma or from Deuteromycetes
infections, Candida and Nocardia are diffusely invasive.
    Infections by species of Aspergillus, Mucorales, and Rhizopus are myceli-
ate opportunists (see Sec. 1.1.2 for a key to the genera). In contrast to the
yeastlike opportunists, the prognosis for patients with these infections is poor.
The majority of the myceliate infections are secondary to other diseases and/
or drug treatment. However, infections may also occur in the absence of
predisposing factors. Interestingly, strains of these fungi produce mycotoxins,
but the role of mycotoxins in diseases produced by these fungi has received
little investigation.

3.12.2.2   Toxomycoses

The second class of pulmonary diseases related to fungi is made up of those
resulting from fungal toxins acting as chemical agents or as allergens. The
absence of fungal viability or growth in vivo distinguishes these mycotic
diseases, which occur after the inhalation of organic dusts contaminated with
fungi. Although Pepys [10] described effects of the dusts as irritant, toxic, or
patently allergenic, these differences could as well be related to dose and site
of deposition as to different mechanisms. Many of these toxomycotic or hypersensitivity pneumonitide responses are named after the occupations through
which they are contacted [11], the most common being farmer's lung (Table 1).
Similar pathologies result from inhalational exposure to chemicals such as
toluene diisocyanate, Teflon, phosgene, and metal fumes. In spite of wide
interest in the toxomycoses, their pathogenesis is not understood. The role of
toxic products in their pathology is largely unexplored.

Table 1   Etiologic Agents in Human Toxomycoses and Alveolitis

Clinical syndrome	Exposure	Antibodies against
Farmer's lung	Moldy crops	Micropolyspora faeni Thermoactinomyces vulgaris T. saccharii
Bagassosis	Moldy bagasse	Same as above
Mushroom worker's lung	Moldy compost	Same as above
Humidifier or air con- ditioner lung	Contaminated ventilation system	Same as above
Maple bark stripper's lung	Moldy bark	Cryptostroma corticale
Malt worker's lung	Moldy malt	Aspergillus spp.
Sequoiosis	Moldy sawdust	Graphium spp.
Cheese washer's lung	Cheese mold	Penicillium spp.
Paprika splitter's lung	Moldy paprika	Mucor stolonifer
Wheat weevil disease	Wheat flour weevils	Sitophilus granarius
Suberosis	Moldy cork	Penicillium spp.
Coffee worker's lung	Coffee bean	Coffee-bean extract
Tobacco grower's lung	Moldy tobacco	Unknown
Wood joiner's lung	Contaminated water	Unknown
Mummy disease	Moldy mummies	Unknown
Bird breeder's lung	Pigeons, parakeets, budgerigars, etc.	Avian serum proteins
Pituitary snuff-taker's lung	Porcine, bovine pituitary	Porcine, bovine pro- teins
Polymer fume fever	Teflon	None
Metal fume fever	Zinc or brass fumes	None
Urethane foam fever	Toluene diisocyanate	None
Cardroom fever	Cotton dust-endotoxin	None
Acute berylliosis	Beryllium	Uncertain

3.12.3   Pulmonary Zones Affected by Fungi

Inhaled fungi, particularly spores, may affect the tracheobronchial tree, the alveoli, or both. The manifestations and duration of disease are highly variable, governed by the intensity of exposure, site of deposition, residence time in the pulmonary zones, and sensitivity of the individual host.

3.12.3.1   Tracheobronchial Tree

Noninfectious fungal exposure to the tracheobronchial tree produces systemic effects of fever, chills, and malaise and pulmonary symptoms of shortness of breath, chest tightness, wheezing, cough, and sputum production. Expiratory airflow is prolonged as measured by a decreased forced expiratory volume during the first second of expiration. An acute exposure frequently elicits an asthmatic response, lasting from a few hours to 2 to 4 days. Prolonged asthma, bronchitis, and eventual debility are the results of chronic exposure.

Occasionally, as fungi grow in mucous plugs, they penetrate the tracheobronchial walls, causing the additional symptom of bleeding. Usually infections develop in a preexisting lesion, e.g., a cavity due to tuberculosis or bacteria or in a bronchiectatic focus. Such fungal balls, which are seen as gray crescents in chest x-rays and are one characteristic type of disease due to Aspergillus, develop slowly, occupy the antecedent lesion, and may cause bleeding.

3.12.3.2   Alveoli

Spores and fungal particles of small size (less than 10 μm) pass through the tracheobronchial tree into the small airways and alveoli to cause localized granuloma, diffuse alveolitis, or infection.

A localized granuloma is characterized by the accumulation of cells around an indigestible particle such as a fungal spore. Recent studies by Boros and Warren [12] have shown only quantitative differences between granulomatous inflammations previously divided into foreign body and hypersensitivity types. The foreign body granulomatous reaction, devoid of immunological elements, consists of macrophages and epithelioid cells. A hypersensitivity granuloma is similar but occurs in the presence of cellular immunity and is correlated with other parameters, skin tests, etc. Hypersensitivity granulomas, e.g., Histoplasma capsulatum, contain lymphocytes and eosinophils in addition to macrophages and epithelioid cells and are usually larger.

The second type of alveolar response to fungi is diffuse toxic or allergic alveolitis without infection. This response is characterized by a wide spectrum of clinical and respiratory features. Radiographic and pathologic observations suggest an interstitial process leading to fibrosis with irreversible pulmonary insufficiency. In severe cases alveolar walls are thickened by infiltration with lymphocytes, plasma cells, and some eosinophils. Diffuse alveolitis occurs in many of the toxomycoses, in which precipitating antibodies to the offending organism are present. Currently, there is much debate on whether the alveolitis seen in the toxomycoses is due to immunological or toxic mechanisms.

Infection, the third alveolar response, occurs when the growth of fungal spores is not checked by cellular and humoral defenses [13]. Disseminated fungal septicemia and multiple organ involvement are the most serious complications of fungal infection in alveoli. They are difficult to recognize, affecting apparently normal as well as immunosuppressed patients, and frequently cause death from toxicity, intravascular clotting, or bleeding. Although fungi grow without constraint, the role of toxins is unexplored.

3.12.4   Mechanisms of Diseases Due
          to Fungi

3.12.4.1   Immune Mechanism

Toxomycoses occur mainly in nonatopic human subjects having humoral, precipitating antibodies in their serum against an agent in the inhaled organic dust.
These precipitating antibodies are helpful in diagnosis, but their role in the
production of alveolitis is uncertain [10]. It has been suggested that insoluble
antigen-antibody complexes attract macrophages and initiate formation of small
granulomata. Presumably, toxic soluble complexes provoke a surrounding
inflammatory reaction and contribute to the pathogenesis of alveolitis. The
main obstacles to inhalation immunization are believed to be the failure to
deposit matter in alveoli, the efficiency of the pulmonary clearance mechanism,
and the inability of alveolar macrophages to present antigens in the manner
necessary to initiate an immune response. Mackaness [14] has suggested that
antigens are conveyed on toxic particles or fungal spores which interfere with
the lungs' clearance mechanism.
    The alternative mechanism of a cellular immune response was suggested
by recent experiments of Richerson [15], in which alveolitis was produced by
aerosol antigen challenge of rabbits showing cellular (delayed) hypersensitivity
reactions but not in animals with humoral immunity. Alveolitis resulting from
a hypersensitivity reaction has also been observed when antigens are part of
an insoluble particle as in Mycobacterium [14] and Schistosoma eggs [12].

3.12.4.2   Toxic Mechanism

In the toxic mechanism for the pathogenesis of the alveolar granuloma and
fibrosis, ingested materials that damage macrophages attract more macrophages and stimulate a fibrogenic reaction. The nature and quantity of toxic
or poorly digestible substances in fungi are crucial to this explanation as are
the cellular mediators or messengers from macrophages. It is generally
assumed that fibrogenesis is a two-stage mechanism wherein interaction of
particles with macrophages results in the release of a factor which stimulates
collagen synthesis by fibroblasts.
    The first stage entails an active factor associated with the particle.
During phagocytosis, the ingested particles fuse with the lysosomes to form
phagolysosomes or secondary lysosomes. Allison [16] has shown that when
nontoxic particles are phagocytosed, the phagolysosomes remain intact.
Several hours after toxic particles as silica dust are taken up, lysosomal
enzymes and peroxidase escape into the surrounding cytoplasm, thereby killing
the phagocytic cell with the consequent release of enzymes that can degrade
extracellular materials. The second stage of collagen synthesis and fibrosis
follows.
    For the application of this hypothesis to the etiology of the toxomycoses,
the accumulation of macrophages is essential. This could occur if the mycotoxin was part of a persistent factor, i.e., a fungal spore. Experimental
models for alveolitis include beryllium, carrageenin, a relatively indigestible
complex polysaccharide derived from seaweed, and bentonite. Labeling

studies [17] have shown that cells which contain phagocytosed persistent irri-
tants were relatively immobile and long-lived. Exceptions were found when
the particles were sufficiently toxic so as to destroy the phagocytic cells
rapidly.

The minimal requirements for this mechanism are an inciting chemical,
persistence of this inciting agent or memory, and cellular damage. Thus,
analyzed this way the immune and toxic mechanisms are similar, the latter
substitutes persistence for memory. This concept would explain the acute
inflammatory reaction as well as the granuloma and fibrosis.

3.12.5   Experimental Evidence for
          Pulmonary Toxicity by
          Mycotoxins

Although few experiments have been published which separate effects of mycotoxins from the responses to spores, the available evidence is provocative.
It is reviewed to synthesize facts from diverse origins and to stimulate further
research.

3.12.5.1   In Vitro Studies

The earliest work showing cellular toxicity of mycotoxins was with cultures of
cells from lung. Legator et al. [18] reported effects of aflatoxin on a heteroploid human embryonic lung cell line, L-132. In the presence of 0.5 ppm of
aflatoxin (Sec. 2.1.2), increases in cell number, DNA, and protein synthesis
ceased after 48 h of culture. These effects were accelerated with higher
concentrations. Although cells remained viable during aflatoxin exposure,
they showed vacuolization, accumulation of cellular debris, and giant cell
formation. It appeared that aflatoxin suppressed DNA synthesis and subsequently inhibited mitosis and protein synthesis.

A comparison of the morphologic effects of several mycotoxins on primary
cell cultures from rat liver, lung, and kidney was reported by Umeda [19].
Liver parenchymal cells were more sensitive than lung or kidney cells to the
hepatocarcinogenic mycotoxins, luteoskyrin (Sec. 2.3.8.1.1), cyclochlorotine
(Sec. 2.3.8.1.2), and aflatoxin B_1 (Sec. 2.1.2). Nivalenol (Sec. 2.5.1.3.2),
rubratoxin B (Sec. 2.3.4), penicillic acid (Sec. 2.3.6), and patulin (Sec. 2.3.5)
had equal effects on all three types of cells. Each mycotoxin produced specific
cytomorphological changes.

Lundborg and Holma [20] devised an experiment to determine whether
spores of an aflatoxin-producing strain of Aspergillus flavus (Sec. 1.2.3.3)
were phagocytized differently from non-aflatoxin-producing strains. Phagocytosis by rabbit macrophages was studied in vitro using electron microscopy.
These cells phagocytized the same number of spores from aflatoxin-producing
and nonproducing strains. The authors did not determine the aflatoxin content
of the spores or whether aflatoxin affects the biocidal capacity or mobility of
macrophages. Aflatoxin has been reported to stabilize lysosomes [21], an

effect which might impair formation of phagocytic vacuoles and permit fungal growth. Sandhu et al. [22] and Merkow et al. [23] found that a poor lysosomal response by cells was correlated with germination and growth of Aspergillus in vivo. However, no comparisons were made of intracellular growth of spores containing aflatoxin and those which did not.

Recent studies that used tracheal organ cultures to bioassay mycotoxins suggested another mechanism for pulmonary disease. Nair et al. [24] showed that when chick tracheal rings were exposed to aflatoxin, cytopathic effects such as exfoliation of ciliated epithelial cells, excess mucin production, disorganized ciliary movement, cellular death, and disintegration occurred. These changes depended on the concentration of toxin and duration of exposure. By allowing the tracheal rings to equilibrate for 24 h, the sensitivity of the bioassay was increased 100-fold [25], to a level of 5 ppb. Tracheal sensitivity of several avian species agreed with their in vivo susceptibility to aflatoxin. Other mycotoxins which had similar effects at higher concentrations were patulin, sterigmatocystin, ochratoxin A (Sec. 2.3.1), sporidesmin (Sec. 2.7.2), and gliotoxin (Sec. 2.3.8.5) [26].

Toxic effects of aflatoxin on tracheal tissue have been observed in our laboratory. Experiments were designed to measure tracheal production of protein. Using the method of Van As [27], isolated rabbit tracheas were inflated with 95% oxygen/5% carbon dioxide and placed in minimum essential medium. After a 30-min equilibration period, the cephalad one-third of the trachea containing accumulated mucus was tied off and removed. This procedure was repeated for the control middle one-third of the trachea and aflatoxin exposed, caudad one-third of trachea. Twice as much tissue, caudad two-thirds, was producing protein for the control measurement as in the aflatoxin experiment, last one-third of trachea. Mucous protein production was 80% higher than control when exposed to 1 to 2 μg of aflatoxin B_1/ml of media. Changes in the consistency of mucus made measurements of volume impractical. Studies using radioactive leucine and acetate in the culture medium indicated that differences in mucous protein could not be attributed to permeability changes. Aflatoxin caused exfoliation and degeneration of the tracheal epithelium, as shown by histological examination. These effects would increase protein content.

3.12.5.2   Animal Studies

To study the effects of inhaled aflatoxin, hamsters and guinea pigs (see also Secs. 3.7.5.1 and 3.7.4.1) were exposed to aerosols of aflatoxin B_1 (Calbiochem, A grade) for 4 h in a cuboidal chamber. The aerosols, which contained approximately 30 μg of toxin in 1% ethanol, were generated with a Collison nebulizer. Four animals were exposed simultaneously and were killed at varying intervals after exposure by intratracheal fixation with osmium tetroxide in fluorocarbon [28]. Exposure of each animal was calculated to be less than one-thousandth of the 30 μg of toxin. Lungs and tracheal tissues were dehydrated and embedded in plastic for light and electron microscopy. Exhaust from the exposure chamber was vented into a detoxifying solution of 5% sodium hypochlorite.

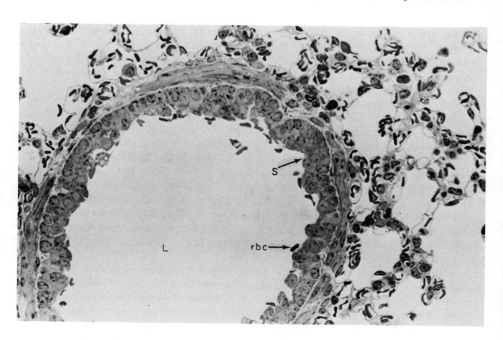

Fig. 1   Columnar bronchial cells of an airway from a guinea pig 6 h after
aerosol exposure of aflatoxin B_1 appear to be stratified (S).   Erythrocytes
(rbc) and exfoliated cells are evident in the lumen (L).   Osmium tetroxide in
fluorocarbon fixation.   Toluidine blue and basic fuchsin stain ($\times$100).

    Within 6 h after the start of aflatoxin exposure there was hemorrhage and
exfoliation of tracheobronchial epithelial cells, sometimes in sheets (Figs. 1
and 2).   The columnar epithelial cells were replaced by round, nonciliated
basilar cells.   This destruction of ciliated and mucus-producing cells would
impair clearance.   Changes were more extensive in the trachea than in smaller
airways, suggesting decreasing deposition as the toxic aerosol moved peripher-
ally.   However, with longer intervals after exposure, peripheral effects
increased.   Two weeks after exposure, ciliated bronchiolar cells appeared in
the alveoli (Fig. 3).   Nonaqueous osmium tetroxide in fluorocarbon fixes
leukocytes and macrophages on top of bronchial cells [28], so displacement of
these or exfoliated cells into alveoli is unlikely.   This metaplasia of ciliated
cells reflects either downgrowth of bronchiolar cells into alveoli or altered
differentiation of alveolar cells.
    These experiments show that acute inhalational exposure to aflatoxin has
destructive effects upon the exposed cells of the respiratory tract and provide
the first experimental evidence of a health hazard of inhaled mycotoxin [29, 30].
The bronchiolar cells in alveoli resemble the effects on hamster lung of
synthetic smog or calcium chromate dust shown by Nettelsheim and Szakal [31].

The tracheal effects are similar to those observed by one of us (K. Kilburn) of other toxic chemicals such as cigarette smoke, gossypol, and paraquat, and benzo(a)pyrene by Port et al. [32]. The damage to tracheobronchial epithelium would impair pulmonary clearance mechanism and would, by prolonging residence time, increase the lung's likelihood of further damage by particles, toxic substances, and microorganisms. This would increase the stress on other defense systems in the lugn.

The carcinogenic effects of aflatoxin on the respiratory tract have been investigated by Dickens et al. [33]. Aflatoxin, dissolved in arachis oil, was administered intratracheally to rats. Mixtures of aflatoxins B_1 and G_1 were given twice weekly in doses of 0.3 mg for 30 weeks. Invasive squamous carcinomas killed three rats at 37, 52, and 62 weeks. The difficulties inherent in producing lung disease with pure substances were emphasized by this study. Coincident tumors of liver, intestine, and kidney in the experimental animals apparently resulted from absorption of the toxin either from the lung or during subsequent passage through the alimentary tract. Recent inhalational studies with benzo(a)pyrene [34], a potent carcinogen found in cigarette smoke, suggests that the pulmonary clearance rapidly eliminates pure substances without their penetration. Use of a surface-active agent (Tween 80) or carrier particle such as ferric oxide facilitates retention, penetration, and increases the rate of disease induction.

A few studies have attempted to assess the role of mycotoxins in experimental aspergillosis. Brewster and Grant [35] studied aflatoxin production in vivo by implants of an aflatoxin-producing strain of A. flavus (Sec. 1.2.3.3). Live and autoclaved mycelia of A. flavus were implanted in the dorsal lymph sac of frogs. Aflatoxin synthesis was shown by the changing ratios of aflatoxin B_1 and G as well as incorporation of [^{14}C]glucose into the excreted toxins. Frogs with live implants excreted 3.5 times as much aflatoxin as those with autoclaved implants. The ratio of aflatoxin B_1 to G excreted by frogs with live implants was 1:2 whereas it was 1:1 for frogs implanted with killed mycelia. These differences may reflect effects of temperature on aflatoxin synthesis [36] in the fungal culture and after implantation and/or metabolism. Before extrapolating to mammals consider that since aflatoxin synthesis is higher at ambient temperatures, little toxin production would be expected from fungi growing in mammals.

After parenteral or inhalational administration of 14 species of Aspergillus to mice, the lungs rapidly disposed of the spores regardless of the pathological potential of the fungal isolate or the route of exposure [13]. Liver and spleen disposed of virulent but not avirulent spores. Aflatoxin was not produced in vivo by an A. flavus strain which produced it in vitro, but neither did fungi grow in mice. In addition, there was no evidence of liver pathology but aflatoxin content of the spores [29] was not given.

Pulmonary alveolar cell hyperplasia (adenomatosis) and diffuse interstitial pneumonia in cattle have been attributed to toxins produced in feeds infected with molds. Fusarium solani (F. javanicum, Sec. 1.4.6.5) was associated with increased respiratory rates, mildly labored breathing, and deterioration if the livestock continued to consume moldy feed. In the terminal stages there

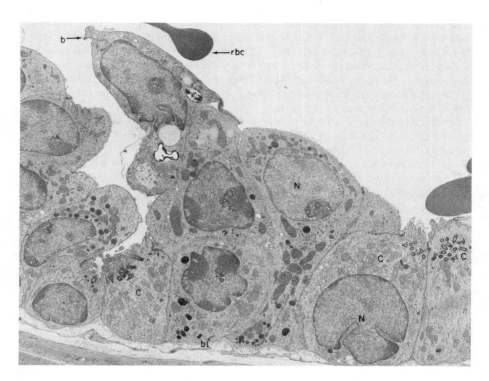

Fig. 2a   Electron micrograph of epithelium from the same animal of Fig. 1.
Nuclei (N) show chromatin clumps and deep indentations of the nuclear
membranes.   Cell membranes appear fuzzy and blebbed (b).   Ciliated cells (C)
attached at the basal lamina (bl) also have blebs on their luminal surfaces.
Uranyl acetate and lead citrate stain (×3300).

was respiratory distress; afflicted animals struggled for air with their necks
extended and mouths opened.   Postmortem examination revealed extensive
pulmonary edema, emphysema, and varying degrees of interstitial pneumonia.
Molds on sweet potatoes, cornstalks, milo, and soybean and peanut hays were
associated with these changes [37-41].

    Wilson et al. [41] investigated the toxicity of extracts of bacteria and
fungi isolated from moldy sweet potatoes.   No toxic symptoms were produced
when the extracts of molds were fed to mice.   However, extracts from the
moldy sweet potatoes were toxic, indicating the toxin was produced in the
tubers in response to fungus infection.   Further studies [42] identified the
lung edema factor as 4-ipomeanol, a furanoterpenoid.   Injection of 1 mg of the
purified compound into mice produced pulmonary edema [43].   This particular
respiratory disease is unique because the toxin was ingested and reached the

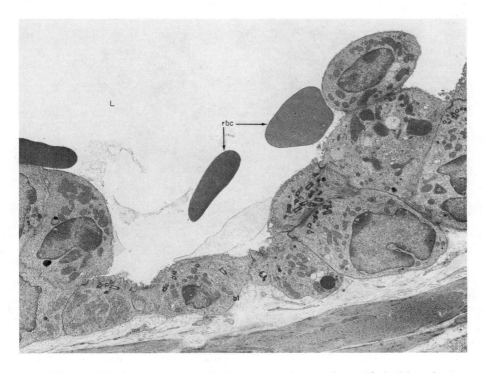

Fig. 2b   Same preparation as (a). Cells attached to the epithelial basal lamina (bl) are thin and attenuated in the middle of this section, while they are stratified on the edges. Erythrocytes (rbc) are seen within the lumen (L) (×3300).

lungs via the splanchnic bed and liver. Finding a heat-stable toxin in marketable sweet potatoes [42] arouses concern for effects on human subjects.

3.12.5.3   Man

The earliest suggestion that human pulmonary disease is produced by mycotoxins is in reports of invasive aspergillosis. Gowing and Hamlin [44] found extensive tissue necrosis around the invading mycelia, suggesting toxic substances were produced by Aspergillus growing in tissue. Enhancement of mycelial growth was attributed to tissue destruction by fungal products. In the absence of tissue necrosis, hyphal penetration seldom reached a depth greater than a few hundred microns [45]. Neutrophils and eosinophils also limited mycotic invasion.

In commenting on disseminated aspergillosis with cutaneous and pulmonary lesions in a 15-year-old girl [46], Eisenberg [47] postulated that hepatic

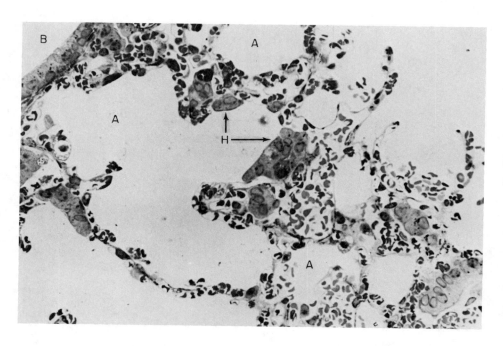

Fig. 3a   This 1-μm section of alveoli from a hamster 2 weeks after exposure to an aerosol of aflatoxin B₁ shows several areas of atypical cells (H) alongside the alveoli (A).   The atypical cells and bronchiolar cells (B) react identically to histological stains.   Toluidine blue and basic fuchsin stain (×100).

necrosis was due to synthesized aflatoxin.   However, aflatoxin production was not studied in vivo or in vitro.

Young et al. [48] showed that pulmonary aspergillosis was not limited to A. fumigatus. Mycological analysis of 39 patients with invasive aspergillosis showed 59% were caused by A. fumigatus, 31% by A. flavus, and 10% by A. glaucus (Sec. 1.2.3.5). Mycotoxin production by fungi isolated from the human respiratory tract has been reported by Fiedoruk-Poplawska [49]. Fungi were cultured on Sabouraud's agar at 22°C. Of 21 Aspergillus isolates, nine produced aflatoxin in vitro, identified as B₁ by thin-layer chromatography of culture extracts in one solvent system and by fluorescence at 365 nm. No other aflatoxins were found.

Isolation of aflatoxin-producing fungi from patients with paranasal aspergillomas has been reported [50]. Fourteen strains of A. flavus cultured on groundnuts and assayed for aflatoxin production by thin-layer chromatography produced aflatoxin in amounts of 80 to 8000 ppb. These strains did not differ morphologically from the usual saprophytic strains of A. flavus. Dissemination

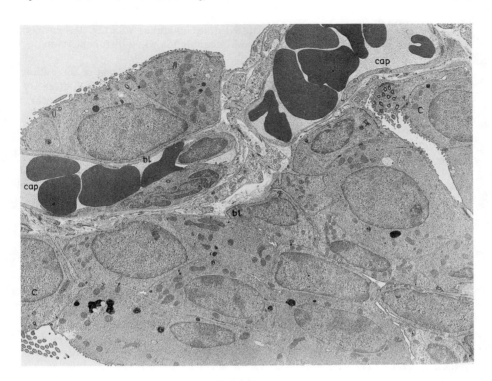

Fig. 3b   Electron micrograph of the section shown in (a). Several ciliated cells (C) attached to alveolar basal lamina (bl) can be seen in the upper region near the capillaries (cap). This has been called bronchiolization of alveoli. There is also an aggregation of atypical cells (right lower foreground) which indicates hyperplasia. Uranyl acetate and lead citrate stain (×3300).

of the infection to the brain resulted in death of one patient. Normal liver functions suggested that aflatoxin, if produced at all, was insufficient to damage the liver.

Production of an aflatoxinlike substance in cultures from 74 patients with aspergillosis has been reported by Poplawska et al. [51]. Serum antibodies against A. fumigatus were present in 65 and absent in 4. Five patients were not tested. Fungi were grown on Sabouraud's agar at 37°C from sputum, pleural effusions, and bronchial washes. Thin-layer chromatography of mycelial extracts with four solvent systems showed a fluorescent substance similar to aflatoxin B_1 in 51% of the strains. The remaining strains, which were obtained from 32 patients, did not produce fluorescent substances. Hemoptysis, hemorrhage, and fever occurred with the same frequency in both

groups of patients. Toxicity of the aflatoxinlike substance was not determined.
Cultural temperatures and chromatography techniques were different from
those previously reported [49].

A patient with pulmonary mycotoxicosis was observed by one of the authors
in August 1973. The patient developed fever, chills, and malaise 6 h after
cleaning a bin of moldy corn. He noted shortness of breath a few hours later
and his diffusing capacity for carbon monoxide, a test for gas transfer, was
reduced. The diagnosis of farmer's lung was supported by precipitan tests.
Culture of the corn showed A. niger (Sec. 1.2.3.7), A. fumigatus, mucor-
mycosis, and a thermophile. A 50-g corn sample, which exhibited bright
blue fluorescence in ultraviolet light, was analyzed for aflatoxin contamination
by quantitative thin-layer chromatography. Aflatoxin B_1 was present at a level
of 6.3 ppb. Observations such as this should be interpreted with care
because the concentration of toxin in the respirable dust from working in the
corn bin may have been much higher than that in the moldy corn. Also,
toxicity would be reduced by the amount of inhaled dust which was rapidly
cleared from the tracheobronchial tree before exerting cytotoxic effects. This
would not affect dust which reaches distal airways or alveoli. Furthermore,
because foreign material cleared from the respiratory tract is swallowed into
the alimentary tract and may be absorbed, liver function should be measured
when airborne aflatoxin exposure is suspected.

In addition to the effects of Aspergillus toxins on pulmonary diseases of
man, a fungal metabolite, oosponol, has been implicated in asthma. The
fungus, Oospora astringenes, was isolated from the air of an asthmatic
patient's room. The compound was absorbed from the culture filtrate onto
active carbon, extracted with methyl acetate, and recrystallized from ethanol
[52]. Further studies [53] identified the asthmatic factor as 3-hydroxymethyl
keton-8-hydroxyisocoumarin. Oosponol caused contractions in a guinea pig
tracheal muscle preparation similar to that induced by barium chloride or an
anaphylactic agent [54]. Oosponol-induced contractions were reversed by
epinephrine, papaverine, khellin, and theophylline. Reserpine, atropine, and
derseril, an antiserotonin, were ineffective. The administration of oosponol
with guinea pig serum albumin was reported to elicit an immune response [55].
These and other studies [5, 56] support an association between asthma and
specific fungal products. Whether the pathogenesis is immunologic or toxic is
still unclear.

3.12.6   Conclusions

Aflatoxins inhaled as aerosols damage avian and mammalian airway cells.
High doses produce hemorrhage, impair pulmonary clearance, and cause
cells to exfoliate. These effects resemble those produced by sulfur dioxide,
oxides of nitrogen, gossypol, paraquat, the vapor phase of cigarette smoke,
and benzo(a)pyrene. A delayed effect in experimental animals is the appearance
of ciliated bronchiolar cells in alveoli which may be an early indicator of
carcinogenic effects in the lung. Human studies suggest that aflatoxin in spores
may contribute to the necrotizing effect of fungi, particularly Aspergillus.

However, no consistent effects of mycotoxins have been demonstrated in man nor have differences been detected between pulmonary disease caused by fungi which produce aflatoxin and those caused by fungi which do not produce it. Thus, we still do not know what roles aflatoxins and other mycotoxins play in human diseases due to inhalation of fungi, but this and other questions concerning the pathogenesis of fungal diseases in the lung need to be answered.

References

1. B. Ramazzini (1964): Diseases of Workers (De Morbis Artificum, 1713). Hafner, New York.
2. P. F. Samsonov (1960): In V. I. Bilai (ed.): Mycotoxicoses of Man and Agricultural Animals. Izd. Akad. Nauk. Ukr. SSR., Kiev, pp. 131-140.
3. F. Kováts, Sr. and B. Bugyi (1968): Occupational Mycotic Diseases of the Lung. Akad. Kiado, Budapest.
4. Anonymous (1969): Occupational pulmonary mycoses and toxomycoses. Can. Med. Assoc. J. 100:583-584.
5. J. Pepys (1969): Hypersensitivity Diseases of the Lungs Due to Fungi and Organic Dusts, Monographs in Allergy, Vol. 4. Karger, Basel.
6. D. Pappagianis (1967): Epidemiological aspects of respiratory mycotic infections. Bacteriol. Rev. 31:25-34.
7. Y. M. Kong and B. H. Levine (1967): Experimentally induced immunity in the mycoses. Bacteriol. Rev. 31:35-53.
8. H. A. Buechner, J. H. Seabury, C. C. Campbell, L. K. Georg, L. Kaufman, and W. Kaplan (1973): The current state of serologic, immunologic and skin tests in the diagnosis of pulmonary mycoses. Report of the Committee on Fungus Diseases and Subcommittee on Criteria for Clinical Diagnosis. American College of Chest Physicians. Chest 63:259-270.
9. D. B. Louria (1967): Deepseated mycotic infections, allergy to fungi and mycotoxins. N. Engl. J. Med. 277:1065-1071, 1126-1134.
10. J. Pepys (1974): Immunologic approaches in pulmonary disease caused by inhaled materials. Ann. N.Y. Acad. Sci. 221:27-37.
11. J. N. Fink (1974): Hypersensitivity pneumonitis due to organic dusts. Clin. Notes Respir. Dis. 13(No. 8):3-9.
12. D. L. Boros and K. S. Warren (1974): Models of granulomatous inflammation. Ann. N.Y. Acad. Sci. 221:331-334.
13. S. Ford and L. Friedman (1967): Experimental study of the pathogenicity of Aspergilli for mice. J. Bacteriol. 94:928-933.
14. G. B. Mackaness (1974): Delayed hypersensitivity in lung diseases. Ann. N.Y. Acad. Sci. 221:312-316.
15. H. B. Richerson (1974): Varieties of acute immunological damage to the rabbit lung. Ann. N.Y. Acad. Sci. 221:340-360.
16. A. C. Allison (1974): Pathogenic effects of inhaled particles and antigens. Ann. N.Y. Acad. Sci. 221:299-308.
17. J. M. Papadimitriou and W. G. Spector (1972): The ultrastructure of high and low turnover inflammatory granulomata. J. Pathol. 106:37-43.

18. M. S. Legator, S. M. Zuffante, and A. R. Harp (1965): Aflatoxin: Effect on cultured heteroploid human embryonic cells. Nature (London) 208:345-347.

19. M. Umeda (1971): Cytomorphological changes of cultured cells from rat liver, kidney and lung induced by several mycotoxins. Jap. J. Exp. Med. 41:195-207.

20. M. Lundborg and B. Holma (1972): In vitro phagocytosis of fungal spores by rabbit lung macrophages. Sabouraudia 10:152-156.

21. A. A. Pokrovsky, L. V. Kravchenko, and V. A. Tutelyan (1972): Effect of aflatoxin on rat liver lysosomes. Toxicon 10:25-30.

22. D. K. Sandhu, R. S. Sandhu, V. N. Damodaren, and H. S. Randhawa (1970): Effect of cortisone on bronchopulmonary aspergillosis in mice exposed to spores of various Aspergillus species. Sabouraudia 8:32-38.

23. L. Merkow, M. Pardo, S. M. Epstein, E. Verney, and H. Sidransky (1968): Lysosomal stability during phagocytosis of Aspergillus flavus spores by alveolar macrophages of cortisone treated mice. Science 160:79-81.

24. K. P. C. Nair, W. M. Colwell, G. T. Edds, and P. T. Cardeilhac (1970): Use of tracheal organ cultures for bioassay of aflatoxins. J. Assoc. Off. Anal. Chem. 53:1258-1263.

25. W. M. Colwell, R. C. Ashley, D. G. Simmons, and P. B. Hamilton (1973): The relative in vitro sensitivity to aflatoxin B_1 of tracheal organ cultures prepared from day-old chickens, ducks, Japanese quail and turkeys. Avian Dis. 17:166-172.

26. P. T. Cardeilhac, K. P. C. Nair, and W. M. Colwell (1972): Tracheal organ cultures for the bioassay of nanogram quantities of mycotoxins. J. Assoc. Off. Anal. Chem. 55:1120-1121.

27. A. Van As (1974): Personal communication.

28. K. H. Kilburn, W. S. Lynn, L. L. Tres, and W. N. McKenzie (1973): Leukocyte recruitment through airway walls by condensed vegetable tannins and quercetin. Lab. Invest. 28:55-59.

29. C. W. Hesseltine, O. L. Shotwell, J. J. Ellis, and R. D. Stubblefield (1966): Aflatoxin formation by Aspergillus flavus. Bacteriol. Rev. 30:795-805.

30. L. A. Goldblatt (1969): Aflatoxin: Scientific Background, Control and Implications. Academic Press, New York.

31. P. Nettelsheim and A. K. Szakal (1972): Morphogenesis of alveolar bronchiolization. Lab. Invest. 26:210-219.

32. C. D. Port, M. C. Henry, D. G. Kaufman, C. C. Harris, and K. V. Ketels (1973): Acute changes in the surface morphology of hamster tracheobronchial epithelium following benzo(α)pyrene and ferric oxide administration. Cancer Res. 33:2498-2506.

33. F. Dickens, H. E. H. Jones, and H. B. Waynforth (1966): Oral, subcutaneous and intratracheal administration of carcinogenic lactones and related substances: The intratracheal administration of cigarette tar in the rat. Br. J. Cancer 20:134-144.

34. V. J. Feron (1972): Respiratory tract tumors in hamsters after intratracheal instillations of benzo(α)pyrene alone and with furfural. Cancer Res. 32:28-36.

35. T. C. Brewster and D. W. Grant (1972): Excretion of aflatoxin by frogs after implantation with Aspergillus flavus. J. Infect. Dis. 125:66-68.

36. D. E. Wright (1968): Toxins produced by fungi. Annu. Rev. Microbiol. 22:269-282.

37. W. Monlux, J. Fitte, G. Kendrick, and H. Dubuisson (1953): Progressive pulmonary adenomatosis in cattle. Southwest. Vet. 6:267-269.

38. W. S. Monlux, P. C. Bennett, and B. W. Kingrey (1955): Pulmonary adenomatosis in Iowa cattle. Iowa Vet. 26:11-13.

39. C. L. Vickers, W. T. Carll, B. W. Bierer, J. B. Thomas, and H. D. Valentine (1960): Pulmonary adenomatosis in South Carolina cattle. J. Am. Vet. Med. Assoc. 137:507-508.

40. J. C. Peckham, F. E. Mitchell, O. H. Jones, and B. Doupnik, Jr. (1972): Atypical interstitial pneumonia in cattle fed moldy sweet potatoes. J. Am. Vet. Med. Assoc. 160:169-172.

41. B. J. Wilson, D. T. C. Yang, and M. R. Boyd (1970): Toxicity of moulddamaged sweet potatoes (Ipomoea batatas). Nature (London) 227:521-522.

42. B. J. Wilson, M. R. Boyd, T. M. Harris, and D. T. C. Yang (1971): A lung oedema factor from mouldy sweet potatoes (Ipomoea batatas). Nature (London) 231:52-53.

43. M. R. Boyd and B. J. Wilson (1972): Isolation and characterization of 4-ipomeanol, a lung-toxic furanoterpenoid produced by sweet potatoes (Ipomoea batatas). J. Agric. Food Chem. 20:428-430.

44. N. F. C. Gowing and I. M. E. Hamlin (1960): Tissue reactions to Aspergillus in cases of Hodgkin's disease and leukaemia. J. Clin. Pathol. 13:396-413.

45. W. St. Clair Symmers (1962): Histopathologic aspects of the pathogenesis of some opportunistic fungal infections, as exemplified in the pathology of aspergillosis and the phycomycetoses. Lab. Invest. 11:1073-1090.

46. B. Castleman and B. U. McNeely (1970): Case records of the Massachusetts General Hospital (Case 44-1970). N. Engl. J. Med. 283:919-927.

47. H. W. Eisenberg (1970): Aspergillosis with aflatoxicosis. N. Engl. J. Med. 283:1348.

48. R. C. Young, A. Jennings, and J. E. Bennett (1972): Species identification of invasive aspergillosis in man. Am. J. Clin. Pathol. 58:554-557.

49. T. Fiedoruk-Poplawska (1971): Zawartosc aflatoksyny B₁ w grzybach wyhodowanych z materialw pochodzacego od ludzi [Content of aflatoxin B₁ in fungi isolated from human materials]. Przegl. Epidemiol. 25:393-397.

50. E. S. Mahgoub (1971): Mycological and serological studies on Aspergillus flavus isolated from paranasal aspergilloma in Sudan. J. Trop. Med. Hyg. 74:162-165.

51. T. Poplawska, H. Halweg, and B. Fronczak (1973): Wytwarzanie substancji aflatoksynopodobnej przez szczepy Aspergillus fumigatus izolowane od chorych na grzybice kropidlakowa pluc [Production of an aflatoxin-like substance by strains of Aspergillus fumigatus from patients with pulmonary aspergillosis]. Gruzlica 41:1034-1042.

52. I. Yamamoto (1961): Studies on the metabolic products of Oospora sp.
    Part I. Isolation and purification of two new compounds and eburicoic
    acid. Agric. Biol. Chem. 25:400-404.
53. I. Yamamoto, K. Nitta, and Y. Yamamoto (1962): Studies on the metabolic
    products of Oospora sp. (Oospora astringenes). Part III. Chemical
    structure of oosponol (0-2). Agric. Biol. Chem. 26:486-493.
54. S. Ohashi, M. Yamaguchi, and Y. Kobayashi (1962): Pharmacological
    studies on anaphylactic and oosponol-induced contraction of tracheal
    muscle preparation from guinea pigs. Proc. Jap. Acad. 38:766-771.
55. K. Uraguchi (1971): Pharmacology of mycotoxins, in: Pharmacology and
    Toxicology of Naturally Occurring Toxins, Vol. II. Pergamon Press,
    New York, pp. 143-299.
56. N. Adiseshan, J. Simpson, and B. Gandevia (1971): The association of
    asthma with Aspergillus and other fungi. Aust. N. Z. J. Med. 4:385-391.

Sharon C. Northup
Kaye H. Kilburn

3.13 Possible Role of Mycotoxins in the Hemic System in Man

Although no direct proof is available to show the effect of mycotoxins on human health, there is good circumstantial evidence to suggest their effect in some disease conditions. Among these is the possible role of mycotoxins in primary carcinoma of the liver as summarized by Alpert and Davidson [1]. These authors reported that statistical analysis of published reports on the incidence of liver cancer in certain parts of Asia and Africa where consumption of foodstuffs contaminated by fungi was high. The excitement produced by the possible carcinogenic effect of mycotoxins in general and aflatoxin in particular led research workers to concentrate on this aspect in relation to the liver. Therefore, when the literature on mycotoxins is reviewed, separate studies on the possible role of mycotoxins on the hemic system are not found, but rather reports on hemic effects, among others.

Before considering the probable effects of mycotoxins on the hemic system of man it would be appropriate to consider such effects in animals, even though it is well known that the effect of mycotoxins on different animal species varies. Christensen et al. [2] have isolated members of the genera Alternaria (see Sec. 1.1.2 for key to genera), Aspergillus (Sec. 1.2.3), Fusarium (Sec. 1.4.6), Penicillium (Sec. 1.3.4), Chaetomium, and Sclerotium from animal feeds and human food, mainly flour, spaghetti products, and pepper. Rats fed with meals inoculated by these fungi showed leukopenia, increased prothrombin time, and hemoglobinuria. Bone marrow studies showed either hypoplasia of all cellular elements or a severe maturation arrest of the myeloid cell series. On the other hand, leukemia in addition to other malignant tumors was produced in mice by fungal extracts from Candida albicans, C. parapsilosis, Scopulariopsis brevicolis, Epidermophyton flocossum, and Microsporum and Trichophyton species, as reported by Blank et al. [3]. Madhavan et al. [4] have reported that the effect on the liver of the rhesus monkey (see also Sec. 3.7.7.1) depended on the dose of aflatoxin (Sec. 2.1.2) and the dietary state of the animal. All animals developed pathological liver changes when given 500 μg/day. Only animals on a low protein diet developed similar changes when the dose was lowered to 100 μg. Control animals in this latter group showed no change. In contrast to these effects Cuthbertson and Laursen [5] denied the effect of aflatoxin on young monkeys that received small doses. Because of this variation in response one is inclined to accept the statement of Barnes [6] that "there is no basis from existing knowledge of the mode of action of aflatoxin and the differing response of various animal species to its acute and chronic effects to conclude how people will respond."

In vitro studies of the effect of aflatoxin on human leukocyte cultures showed that aflatoxin B_1 (Sec. 2.1.2.2) preparations inhibited mitosis and caused breakage and translocations in the chromosomes [7].

A statistically significant and experimentally interesting study was that of Kinosita et al. [8] in Japan. Vital statistics in Japan showed that the main causes of death were apoplexy, neoplasms, and heart disease. These authors suspected fermented foodstuff, a major type of food in Japan, as the possible cause due to its extensive consumption. A. flavus (Sec. 1.2.3.3) and A. oryzae as well as some 35 other strains of fungi were isolated from such foods. To see their effect in animals, culture filtrates were prepared and inoculated into mice. At postmortem, myocardial hemorrhages were produced by a strain of A. glaucus (Sec. 1.2.3.5), A. candidus (Sec. 1.2.3.1), and another Aspergillus species as well as Pestalotia species. This is a very interesting finding, but whether or not the same effect takes place in man cannot be easily confirmed. Many fungi in this study produced kojic acid and β-nitropropionic acid but none produced aflatoxin.

A more definite but rather rare disease condition in which consumption by man of cereals contaminated with toxic fungi resulted in toxic effects is that of alimentary toxic aleukia or leukopenia (ATA), reported from Russia [9]. (See Secs. 3.9.2.2 and 3.10 in this text.) Patients who acquired this condition developed leukopenia, agranulocytosis, and exhaustion of the bone marrow, in addition to other intestinal symptoms. Joffe [9] fed animals similar moldy cereals and managed to obtain effects very similar to those described in man. He therefore postulated that ATA was caused by toxins produced by fungi growing in overwintered grain. Among these fungi were species of the genera (see Sec. 1.1.2) Fusarium (Sec. 1.4.6), Cladosporium, Alternaria, Penicillium (Sec. 1.3.4), and Mucor. Toxic strains grew at temperatures as low as $0°C$, but nontoxic strains did not. No toxin was produced at 23 to $25°C$. In addition, Joffe and Brout [10] found that their species of Fusarium produced a toxin called fusariogenin, and Cladosporium species (see Sec. 1.1.2 for key to genus) produced epicladosporic acid and fagiocladosporic acid. These toxic substances produced bone marrow aplasia in laboratory animals.

All these findings, whether in man or animals, led research workers to speculate that mycotoxins are more likely than other factors to be the cause of certain pathological manifestations. Alexandrowicz et al. [11] have suggested the possible role of metabolic fungal products in the etiology of some aplastic or proliferative blood diseases. In their investigation of homes of patients in different localities in Poland, these workers noted two relevant findings:
(a) A variety of fungi known to produce aflatoxins B_1, B_2, G_1, G_2 (Sec. 2.1.2), fusariogenin, epicladosporic acid, fagiocladosporic acid, and other undefined toxic substances were found in the dwellings of patients with leukemia.
(b) Contrary to this, either no toxic fungi or only few were present in the houses of healthy controls.

Another human condition for which there is some evidence of a causative mycotoxin is Reye's disease (see also Sec. 3.9.3). This is a disease entity of childhood characterized by encephalopathy and fatty degeneration of the viscera, particularly the liver. The syndrome was described originally by Reye et al.

[12] and further elaborated on by Becroft [13]. According to these workers it is an acute and highly fatal illness. Disturbed consciousness, vomiting, disturbed respiratory rhythm, and altered muscle reflexes were seen. In addition to changes in levels of serum glutamic oxaloacetic transaminase (SGOT) and serum glutamic pyruvic transaminase (SGPT), which are due to pathological changes in the liver, there is also hypoglycemia, leukocytosis, mainly in neutrophils, low prothrombin index, and low serum carbon dioxide content. Fatty infiltration of the myocardium was noted less frequently. Sometimes hemorrhages in the intestinal canal and petechial hemorrhages in the skin were seen. Reactive changes appeared in lymphoid tissue. Lymphoid follicles showed large pale central areas in which there were lipid-filled histiocytes and much karyorrhexis.

Becroft and Webster [14] hypothesized that Reye's syndrome could be due to a mycotoxin since the liver lesions were toxic in nature and simulated those produced experimentally in animals given aflatoxin. They theorized that children, the only age group in which the disease is encountered so far, are unlikely to acquire toxins except in foodstuffs. They quoted Shank et al. [15] who identified aflatoxin in the organs of children from Thailand who had a syndrome similar to Reye's disease and had previously consumed rice contaminated with fungi. Becroft and Webster later extracted livers from two patients with Reye's disease and tested these extracts by means of chromatography against standard aflatoxin solutions. Although these extracts gave fluorescent bands, they were not identical with those of aflatoxin.

There are cases of disseminated or septicemic aspergillosis as reported by Vedder and Schor [16] and Mahgoub and El Hassan [17], where the disease was primarily in the lungs, but subcutaneous and skin nodules were also present. Infection was caused by A. flavus. Histological sections showed the picture of both acute and chronic arteritis. No fungal invasion of the arteries was noted. Although one is inclined to think that the basis of this reaction may be immunological, the possibility of fungal metabolites cannot be excluded. There is no definite evidence to show that toxic metabolites are released from such massive Aspergillus growths in the body. However, Eisenberg, commenting on a case of hepatitis and Aspergillus septicemia (Case 44-1970), pointed to the possibility of a mycotoxin [18, 19]. This patient developed leukocytosis with lymphopenia, prolonged prothrombin time, and hemorrhages before death. There was evidence of toxic hepatitis. Although the various symptoms in this case might have been due to liver disease as such, the possibility of a toxin could not be excluded. Eisenberg felt it conceivable that disseminated aspergillosis could result in mycosynthesis of levels of aflatoxin sufficient to cause hepatic necrosis [18].

In addition to aflatoxin, A. flavus produces another toxic substance known as kojic acid, which proved to be toxic to human leukocytes and cardiotoxic in chick embryos [20]. Kojic acid is also produced by some members of the genus Penicillium. Toxicity of a third substance, β-nitropropionic acid, which is produced by A. flavus, is related to a vasodilatory action on peripheral vessels and may therefore result in slowing the rate of blood circulation.

Study of histological sections from paranasal Aspergillus granuloma cases in Sudan due to A. flavus also revealed vascular changes in about one-third of

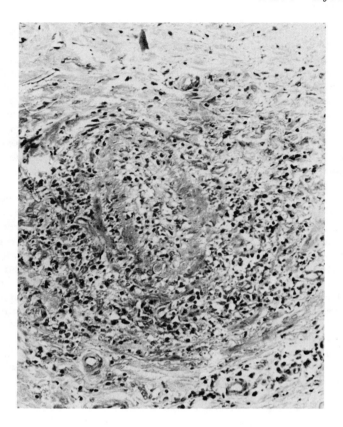

Fig. 1   Acute arteritis in a case of paranasal _Aspergillus_ granuloma (H&E, ×200).

them [21]. The changes were those of either acute or chronic endarteritis (Fig. 1). The latter type showed (a) onionlike perivascular fibrosis (Fig. 2), (b) perivascular cuff formation by lymphocytes and plasma cells, and (c) fibrosis with intimal proliferation (Fig. 3). Hematological examination of blood samples from these patients did not show abnormal values in white cell count. We know that _A_. _flavus_ strains isolated from these lesions produced both aflatoxin and kojic acid in the laboratory and we also demonstrated that all sera from paranasal granuloma patients gave precipitin lines with _A_. _flavus_ antigens [22]. Although we cannot tell if toxic fungal metabolites play some role in these vascular changes, the possibility is great. One wonders if conditions in the human body are conducive to toxin production, particularly the high body temperature. The optimum temperature for aflatoxin production varies with the fungal species but is generally between 25 and 30°C. At high temperatures there is a decline in toxin production [23, 24].

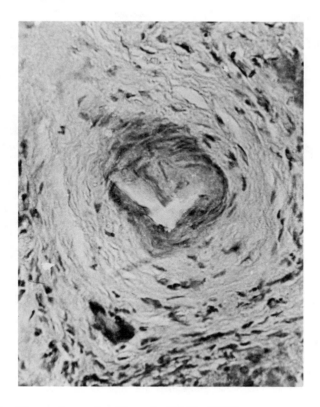

Fig. 2   Onionlike perivascular fibrosis (H&E, ×200).

In summary, it seems reasonable to assume that man is exposed to mycotoxins as much as other animal species, but he may not necessarily respond in
the same manner. The source of mycotoxins is more likely to be consumption
of contaminated foodstuffs than multiplication of fungi in the human body.

Deleterious experimental effects of mycotoxins on the hemic system of
animals are conclusive. Such effects vary from anaplastic to proliferative
changes in white blood cells as well as myocardial hemorrhages. The role of
mycotoxins in leukemia is highly speculative. The role in Reye's syndrome and
paranasal Aspergillus granuloma is highly probable, and in alimentary toxic
aleukia (ATA) is more definitive. At present it is not completely known whether
aflatoxin alone or other toxic substances produce these effects. From the
foregoing information it is worth noting that not only A. flavus, but other
Aspergillus species can produce toxic substances. Species of Fusarium, Mucor,
Cladosporium, Penicillium, and Alternaria also produce toxic products. The
more one reviews this area of research, the more one finds the need for definite
information which can only be obtained through further experimental work.

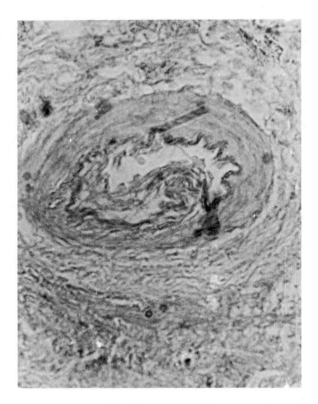

Fig. 3   Fibrous intima proliferation (H&E, ×200).

Acknowledgment

I would like to express my gratitude to Dr. Veres of the pathology department who helped in the selection of the histopathological figures.

References

1. M. E. Alpert and C. S. Davidson (1969): Mycotoxins — A possible cause of primary carcinoma of the liver. Editorial. Am. J. Med. 46:325-329.
2. C. M. Christensen, G. H. Nelson, C. J. Mirocha, and Fernbates (1968): Toxicity to experimental animals of 943 isolates of fungi. Cancer Res. 28:2293-2295.
3. F. Blank, O. Chin, G. Just, D. R. Meranze, M. B. Shimkin, and R. Wieder (1968): Carcinogens from fungi pathogenic for man. Cancer Res. 28:2276-2281.

4. K. Madhavan, K. S. Rao, and P. G. Tulpule (1965): Effect of dietary protein level on susceptibility of monkeys to aflatoxin liver injury. Indian J. Med. Res. 53:984-989.

5. W. F. J. Cuthbertson and A. C. Laursen (1967): Effect of groundnut meal containing aflatoxin on cynomolgus monkeys. Br. J. Nutr. 21:893-908.

6. J. M. Barnes (1970): Aflatoxin as a health hazard. J. Appl. Bacteriol. 33:285-298.

7. D. A. Dolimpio, C. Jacobson, and M. Legator (1968): Effect of aflatoxin on human leucocytes. Proc. Soc. Exp. Biol. Med. 127:559-562.

8. R. Kinosita, T. Ishiko, S. Sugiyama, T. Seto, S. Igarasi, and I. E. Goeta (1968): Mycotoxins in fermented food. Cancer Res. 28:2296-2311.

9. A. Z. Joffe (1964): Toxin production by cereal fungi causing toxic alimentary aleukia in man, in G. N. Wogan (ed.): Mycotoxin in Foodstuffs. M.I.T. Press, Cambridge, Massachusetts.

10. A. Z. Joffe and S. Y. Brout (1966): Soil and kernel mycoflora of groundnut field in Israel. Mycologia 58:629-640.

11. J. Alexandrowicz, B. Smyk, M. Czachor, and Z. Schiffer (1970): Mycotoxins in aplastic and proliferative blood disease. Lancet 1:43.

12. R. D. K. Reye, G. Morgan, and J. Baral (1963): Encephalopathy and fatty degeneration of the viscera. A disease entity in childhood. Lancet 2:1061.

13. D. M. O. Becroft (1966): Syndrome of encephalopathy and fatty degeneration of viscera in New Zealand children. Br. Med. J. 2:135-140.

14. D. M. O. Becroft and D. R. Webster (1972): Aflatoxins and Reye's disease. Br. Med. J. 4:117.

15. R. C. Shank, C. H. Bourgeois, N. Keschamras, and Mol. P. Chandav (1971): Food Cosmetol. Toxicol. 9:501.

16. J. S. Vedder and W. F. Schor (1969): Primary disseminated pulmonary aspergillosis with metastatic skin nodules. J. Am. Med. Ass., 209:1191.

17. E. S. Mahgoub and A. M. El Hassan (1972): Pulmonary aspergillosis caused by Aspergillus flavus. Thorax 27:33-37.

18. H. W. Eisenberg (1970): Aspergillosis with aflatoxicosis. N. Engl. J. Med. 283:1348.

19. Case Records of the Massachusetts General Hospital (Case 44) (1970): N. Engl. J. Med. 283:919-927.

20. B. J. Wilson (1966): Toxins other than aflatoxins produced by Aspergillus flavus. Bacteriol. Rev. 30:478-484.

21. B. Veress, O. S. Malik, A. A. Eltayeb, E. Daoud, E. S. Mahgoub, and A. M. El Hassan (1973): Further observations on the primary paranasal aspergillus granuloma in the Sudan. A morphological study of 46 cases. Am. J. Trop. Med. Hyg. 22:765-772.

22. E. S. Mahgoub (1971): Mycological and serological studies on Aspergillus flavus isolated from paranasal aspergilloma in Sudan. J. Trop. Med. Hyg. 74:162-165.

23. A. Ciegler and E. B. Lillehoj (1968): Mycotoxins. Adv. Appl. Microbiol. 10:155-219.

24. H. F. Kraybill (1969): Review, the toxicology and epidemiology of mycotoxins. Trop. Geogr. Med. 21:1-18.

El-Sheikh Mahgoub

PART 4

EFFECTS OF MYCOTOXINS ON HIGHER PLANTS, ALGAE, FUNGI, AND BACTERIA

4.1 Introduction

Mycotoxins are, according to a generally accepted definition, substances that are produced by fungi on foods and feeds and that can bring about specific intoxication symptoms in animals and very probably also in man. Although many mycotoxins may be toxic to plants they are never involved in the development of plant diseases [1]. Because of this fact the mycotoxins may be distinguished from the phytotoxins which are metabolites of phytopathogenic fungi which intoxicate first of all the host plants. A third group of fungal metabolites, the antibiotics, may also be harmful to higher plants and to animals but are mainly toxic to bacteria. So the definition of the three terms overlaps, and is a matter of convenience [1].

In the following the effects of different mycotoxins on higher plants, algae, fungi, and bacteria are described.

4.2 Aflatoxins

4.2.1 Effects on Higher Plants

4.2.1.1 Effects on the Germination of Seeds

Schoental and White [2] observed a considerable reduction of the germination of seeds of Lepidium sativum after treatment with 10 to 50 μg/ml of crude aflatoxin (Sec. 2.1.2). The synthesis of chlorophyll was suppressed totally by 10 μg toxin/ml. Using pure aflatoxin B_1 (Sec. 2.1.2.2) Reiss [3] found no differences in the percentage of germinated Lepidium sativum seeds under the influence of as high as 100 μg toxin/ml; the development of the radicula and especially that of the hypocotyl of the seedlings were markedly inhibited by 10 and 100 μg toxin/ml. A concentration of 1 μg toxin/ml enhanced the formation of both organs (Table 1). Crisan [4] obtained similar results in his study on the influence of an aflatoxin mixture on germination and growth of 30 cultivars of lettuce (Lactuca sativa). The germination itself was not influenced even by 1000 μg toxin/ml but the elongation of the hypocotyl was inhibited by 46 to 68% at a toxin concentration of 100 μg/ml.

The germination of seeds of 19 species of Cruciferae was not affected by 100 μg aflatoxin B_1/ml, but the formation of the hypocotyl and radicula was inhibited to a high degree. Lepidium sativum was the most susceptible plant [5]. Aflatoxin B_1 (Sec. 2.1.2.2) and crude aflatoxins inhibited chlorophyll formation and the germination of the cowpea. These effects were reduced by β-indolylacetic acid [5a].

119

Table 1   Influence of Mycotoxins on the Development of the Hypocotyl (a) and the Radicula
(b) of <u>Lepidium</u> <u>sativum</u>

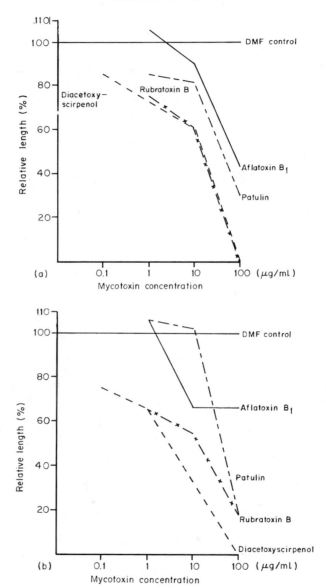

Table 2   Influence of Mycotoxins on the Action of
β-Indolylacetic Acid in Sections of Pea
Internodes

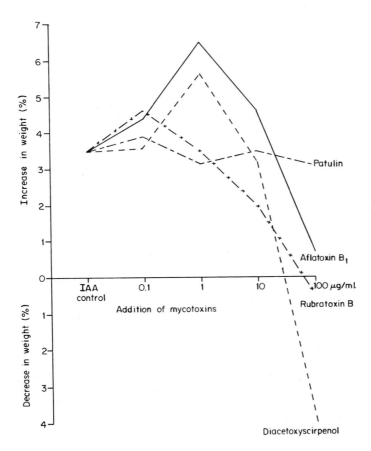

4.2.1.2   Effects on β-Indolylacetic Acid

Aflatoxins are coumarin derivatives with a lactone grouping. Coumarin (Sec.
2.2) and unsaturated lactones are known to be synergists of β-indolylacetic acid
(IAA) in low concentrations and antagonists in higher. This was confirmed for
aflatoxin B₁ using stem sections of <u>Pisum</u> <u>sativum</u> (Table 2) [6].

4.2.1.3   Effects on Nucleic Acids,
           Enzymes, and Other Proteins

Gibberellic acid stimulates the formation of the enzymes lipase and α-amylase
in germinating cottonseed (Gossypium hirsutum) [7]. When gibberellic acid-
treated distal halves of seeds were exposed to aflatoxin, α-amylase and lipase
activities were strongly inhibited. It was suggested that the mycotoxin is an
inhibitor of protein synthesis [7, 8]. Truelove et al. [9] continued this work
and studied the effect of aflatoxin B_1 on the uptake of [^{14}C]leucine and its
incorporation into protein by disks of cotyledons of cucumber (Cucumis sativa).
Preincubation with 5 μg toxin/ml for 6 h reduced the uptake of [^{14}C]leucine by
about 50% without affecting its incorporation into protein. Extending the pre-
treatment to 12 h or increasing the toxin concentration reduced the incorporation
of the amino acid into protein.

Asahi et al. [10] and Uritani et al. [11] investigated the effect of aflatoxin
B_1 on the metabolic activity of root tissues of sweet potato (Ipomoea batatas).
The toxin inhibited the reproduction of mitochondria but did not affect the
synthesis of peroxidase, cytochrome oxidase, and succinate dehydrogenase.
The authors suppose that aflatoxin B_1 inhibits the replication of mitochondrial
DNA but has no effect on the protein synthesis itself.

According to the study of Dashek and Llewellyn [11a] aflatoxin B_1 interferes with protein synthesis in germinating pollen of Lilium longiflorum.

4.2.1.4   Other Physiological Effects

Wilting and dessication of several field crop plants (e.g., groundnut, tomato,
maize, Leguminosae) after treatment with solutions in which toxicogenic
strains of Aspergillus flavus (Sec. 1.2.3.3) had been cultivated were described
by Joffe [12]. Both 300 and 100 ppm crude aflatoxin stopped the growth of
branches of Caralluma frerei (succulent Asclepiadaceae) [13]. Treatment of
peanut seedlings with pure aflatoxin B_1 as well as with extracts of liquid
cultures of Aspergillus flavus resulted in extreme reduction of growth (Fig. 1)
[14, 15]. Application by the roots was more effective than application to the
cotyledons.

4.2.1.5   Effects on Cellular
           Organization

Lilly [16] treated roots of Vicia faba seedlings with a crude aflatoxin mixture
and squashed the preparations after staining with the Feulgen technique.
There was a significant increase in the number of abnormal anaphases and a
considerable inhibition of mitosis. Chromosomal aberrations (fragmentation
with bridges) were observed. Similar damage was observed after treatment of
roots of Allium cepa with aflatoxin B_1 solutions [17, 18]. Chromosome
bridges, C-mitoses, and a reduction of the mitotic index were the main effects
(Fig. 2).

Fig. 1   Reduced growth of peanut seedlings after application of pure aflatoxin
B_1 through the roots (from left to right, 0, 1, 5, 10 ppm toxin).   Photograph
through the courtesy of Professor F. Grossmann.

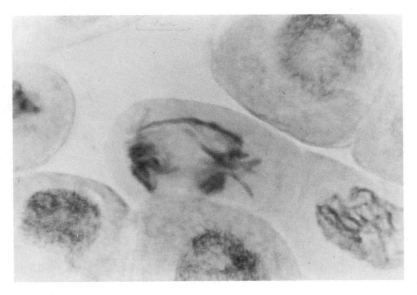

Fig. 2   Anaphase bridge in <u>Allium cepa</u> root meristem cell after cultivation
in 100 μg aflatoxin B_1 /ml for 46 h.

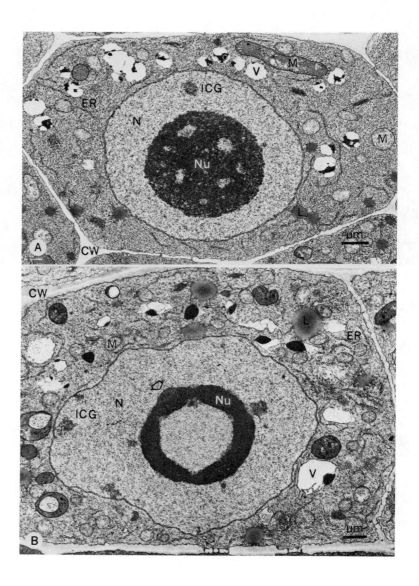

Fig. 3   Ultrastructure of root cells of <u>Lepidium</u> <u>sativum</u>. (A) Control,
(B) treated with 100 μg aflatoxin B_1 /ml. Abbreviations: CW, cell wall; ER,
endoplasmic reticulum, IGG, interchromatin granules; L, lipid bodies; M,
mitochondria; N, nucleus; Nu, nucleolus; V, vacuole; arrow, light nuclear
cap. Reprinted from Crisan [5, p. 997] through the courtesy of the American
Society for Microbiology.

Electron microscopic investigations of aflatoxin-treated root cells of Lepidium sativum revealed cytological changes similar to those known in aflatoxin-treated animal cells [5]. Most prominent was the occurrence of ring-shaped nucleoli with distinct nucleolar caps (Fig. 3), an increased number of lipid bodies, mitochondria with enlarged cisternae, degranulation of the endoplasmic reticulum, enlarged and irregularly shaped nuclei, and increased number of interchromatin granules. It is assumed that aflatoxin B_1 acts as an antimetabolite that binds to DNA and interferes with RNA and protein bio-synthesis.

The most prominent ultrastructural changes in meristem cells of the root of Allium sativum were a reduction of the number of ribosomes and an increase of the dictyosomes [18a].

Slowatizky et al. [19] treated etiolated leaves of Zea mays with solutions of aflatoxin and exposed the leaves to fluorescent light. As a result the leaves turned green while the spots on which aflatoxin had been placed remained yellow. The electron microscopic picture showed an inhibition of the formation of the grana in the chloroplasts. As little as 0.25 μg aflatoxin had this effect and therefore a simple but sensitive bioassay for the detection of afla-toxin in milk was developed [20]. A similar disintegration of the chloroplasts could be observed in Nasturtium officinale [18a].

4.2.2   Effects on Algae

Concentrations of 4 μg aflatoxin B_1/ml in liquid culture and 12.5 μg/disk in a disk assay procedure inhibited the growth of Chlorella pyrenoidosa [21]. The aflatoxins B_2, G_1, and G_2 showed no effects [22]. Euglena gracilis was far less sensitive [21].

4.2.3   Effects on Fungi

Lillehoj et al. [23] were the first who tested the influence of aflatoxins on molds. Aflatoxin B_1 inhibited the growth of Aspergillus flavus (four isolates), A. awamori, Penicillium chrysogenum, and P. duclauxi. In Mucor hiemalis 10 ppm B_1 inhibited the cytochemically detectable activity of the enzymes succinate dehydrogenase, alcohol dehydrogenase, L-isocitrate: NAD oxido-reductase, and acid phosphatase [24]. A crude mixture of aflatoxins in a concentration of 100 μg/filter disk (9 mm) inhibited the sporulation of Aspergillus niger (Sec. 1.2.3.7), Penicillium expansum (Sec. 1.3.4.8), Cladosporium herbarum, Mucor hiemalis, Rhizopus nigricans (see Sec. 1.1.2 for key to genera), and Thamnidium elegans [25]. The germination hyphae of the latter fungus showed distinct deformations when cultivated under the influence of 100 μg aflatoxin B_1/ml [26].

Abnormalities in the vegetative mycelium of Aspergillus flavus under the influence of aflatoxin were described by Wildman [27]. Jacquet and Boutibonnes [28] found giant cells in A. parasiticus (Sec. 1.2.3.3) after treatment with 50 μg aflatoxin/ml. Electron microscopy of these giant cells revealed damaged mitochondria, pinocytosis, and changes of the cell wall.

The fermentation activity of Saccharomyces cerevisiae is influenced by aflatoxin B_1: Concentrations of 1 μg/ml or less supported CO_2 production whereas 10 and 100 μg/ml exerted an inhibitory effect. Enzyme cytochemical investigations revealed a deactivation of the alcohol dehydrogenase of the yeast by 100 μg toxin/ml and an increase of activity by 0.1 to 1 μg/ml [29].

Extensive studies by Ong [30-33] demonstrated a mutagenic activity of the aflatoxins in the ad-3 system of Neurospora crassa. G_1 was less mutagenic than B_1. This compound was mutagenic in growing vegetative cultures of N. crassa but not in resting conidia. Aflatoxin B_1 gave a 177-fold increase over the spontaneous mutation frequency and must be considered as a potent mutagen for this mold.

Aflatoxins may even be degraded by fungi. Dactylium dendroides, Absidia repens, and Mucor griseo-cyanus transformed about 50 to 60% of aflatoxin B_1 in shake culture to a new fluorescent-blue compound [33a]. Cole and collaborators found that species of Rhizopus (commonly associated with peanuts in Georgia) are able to degrade aflatoxin G_1 (Sec. 2.1.2) to aflatoxin B_3 (parasiticol) [34] and aflatoxin B_1 to hydroxylated stereoisomers [35].

4.2.4   Effects on Bacteria

Burmeister and Hesseltine [36] conducted the first investigations on the influence of aflatoxins on bacteria. Only gram-positive, spore-forming bacilli were found to be sensitive and Bacillus megaterium and B. brevis were most susceptible. Continuing studies demonstrated that 4 × 10^9 B. megaterium/ml took up approximately 7 μg aflatoxin B_1/ml, which remained tightly bound [37]. B. megaterium grown in the presence of the toxin showed aberrant elongation of the cells due to an inhibition of the mechanism of cell division. Since the mechanism of cytokinesis in bacteria probably involves an interaction between nucleus and cell membrane the induction of aberrant forms might be due to an interference of the aflatoxin B_1 molecule with nucleic acid synthesis [37]. Filamentous forms of B. megaterium under the influence of aflatoxin B_1 were also described by Beuchat and Lechowich [38]. The treated cells contained unusually large and numerous deposits of poly-β-hydroxybutyric acid and a decreased number of mesosomes. The transfer of aberrant cells to nontoxic agar yielded colonies with daughter cells morphologically indistinguishable from untreated cells. Chemical analysis revealed an increase in protein, DNA, and RNA in B. megaterium cells after cultivation in Trypticase Soy Broth containing 3.8 μg aflatoxin B_1/ml. Differences in cell wall composition of control and test cells were not found. Aflatoxin B_1 inhibited the sporulation rate of B. megaterium [39]. The cell wall is not required for the absorption of aflatoxin by B. megaterium since an L-form was inhibited by the same toxin concentration as the intact Bacillus [40]. Aflatoxin B_1 induced lysis in a lysogenic strain of B. megaterium [41].

The extreme sensitivity of B. megaterium toward aflatoxins led to the development of biotests by which as low as 1 μg B_1 and 4 to 8 μg G_1 can be detected in a modified filter paper technique [42-46].

Lillehoj and co-workers studied the toxic effects of aflatoxins toward
Flavobacterium aurantiacum. A concentration of 5 ppm B_1 inhibited the growth
and caused the development of aberrant filamentous forms. The nature of the
inhibition suggested that the toxin interferes with cell wall synthesis [47].
Incubation of 10^{11} resting cells/ml with 7 μg aflatoxin B_1/ml facilitated a
complete toxin removal during a 4-h period [47]. In F. aurantiacum the
synthesis of DNA was more sensitive to aflatoxin than the synthesis of RNA
[48]. Aflatoxin G_1 proved to be less toxic to this organism than B_1. Only a
few cells showed aberrant morphological forms. During an incubation period
of 4 h, 330 μg G_1 was removed by 10^{13} resting cells [49]. The growth of F.
aurantiacum was likewise inhibited in the presence of aflatoxin M_1 but this
compound was less toxic than B_1 and G_1 [50].

Beside B. megaterium and F. aurantiacum other bacteria may be intoxica-
ted by the aflatoxins. Filamentous forms and a decrease of the activity of DNA
polymerase and reduction of the synthesis of DNA were observed in Escherichia
coli under the influence of aflatoxin B_1 [51]. From 10 species of bacteria
Bacillus mycoides and Brevibacterium species proved to be especially sensitive
toward aflatoxin B_1 [52]. Under the influence of 50 μg aflatoxin Bacillus
thuringiensis [53, 54] and B. stearothermophilus [28, 55] formed abnormal
irregularly shaped cells with strongly granulated cytoplasm and only a weak
stainability. The strong susceptibility of B. stearothermophilus toward mycotoxins led to the development of a sensitive and simple bioassay procedure for
aflatoxin B_1 and other mycotoxins [56]. A mixture of aflatoxins B and G
inhibited the growth of gram-positive bacteria (Bacillaceae and Coryne-
bacteriaceae) and induced the formation of swollen cells in these species [57].
Species of Streptococcus and Nocardia were also sensitive toward aflatoxin B_1
[58]. Biochemical studies revealed a decrease in the oxygen consumption in
strains of Pseudomonas aeruginosa, Proteus vulgaris, Staphylococcus aureus,
and E. coli by 10 μg aflatoxin B_1/ml [59]. In Bacillus cereus and B. licheniformis the formation of penicillinase and α-glucosidase was partially blocked
by aflatoxin B_1 [60].

4.3   Patulin

Patulin (Sec. 2.3.5) is produced by several species of Aspergillus (Sec. 1.2)
and Penicillium (Sec. 1.3). Because of its pronounced antibacterial activity
this compound was supposed to be a potent antibiotic, but its toxicity to animals
is too high so it cannot be used as a therapeutic agent.

4.3.1   Effects on Higher Plants

Norstadt and McCalla extensively studied the cause for decreased yields and
abnormal appearance of crops during stubble-mulch farming in Nebraska. They
isolated Penicillium urticae (Sec. 1.3.4.13) from surface-tilled plots and
identified the toxic principle as patulin [61, 62]. Artificial additions of patulin

to wheat plants reduced the yields of straw, root, and grain and seed germination. The affected plants showed necrosis, narrowed and shortened leaves,
leaf-tip burning, reduction in diameter and length of the stem, shortening of
the first internode, and chlorosis [63, 64]. Increasing doses of patulin
affected the plants more strongly in sand than in soil. This observation is
suggested to be the result of an inactivation of the toxin by soil microorganisms
[64]. Even a single treatment of roots of wheat with 100 μg patulin/ml led to a
reduction in internodal elongation, floret number, seed weight, and seed
number [65]. It is assumed that at least the reduction of internode length is
due to an interference of patulin with sulfhydryl groups of essential enzymes.
Solutions of patulin in a concentration of 1:4000 reduced the germination of
wheat seeds to 53% [66].

In his study on soil sickness in apple nurseries Börner isolated a strain of
Penicillium expansum (Sec. 1.3.4.8) from soil that was able to produce patulin
on root bark or foliage. The toxin was also found in soils with normal microflora [67]. Root residues of apple, sour cherry, and wheat straw resulted in
an especially strong patulin formation whereas high concentrations of inorganic
nutrients decreased the toxin yields [68].

Branches of tomato (Solanum lycopersicum) which had been cultivated in
solutions of patulin showed deformations on leaf stalks and a brownish
coloration of the vessels; higher doses induced wilting [69]. The vessels,
phloem, and bark tissue of the shoots were most damaged [69a]. Fifty per-
cent of all sulfhydryl compounds (and therefore important enzymes) were
inactivated by patulin in tomato branches [69]. Patulin in a concentration of
60 μg/ml was lethal to tomato plants and produced very severe wilting and
slight stunting and leaf necrosis, but no bleaching [70].

Patulin inhibited the development of hypocotyl and radicula in Lepidium
sativum [71] (Table 1) and the germination of pea seeds [70a]. An interference with β-indolylacetic acid in internodes of Pisum sativum could not be
seen [71] (Table 2).

Several investigators demonstrated that patulin acts as a cytotoxic agent.
In Allium cepa a reduction of the mitotic index [72, 73], extreme vacuolization,
and the appearance of binucleate cells [74] were the most prominent symptoms
of intoxication.

Wilting phenomena in Malva rotundifolia, Brassica oleracea var. botrytis,
and Pisum sativum after treatment with culture filtrates of a toxicogenic strain
of Penicillium expansum were very probably due to the influence of patulin
[75, 76].

4.3.2   Effects on Fungi

The antifungal activity of patulin has been established by several studies. The
phytopathogenic fungi Ustilago tritici [66], Pythium species, and Phytophthora
erythroseptica [77, 78] as well as Ascochyta pisi [70a] proved to be very
sensitive. The growth of Aspergillus niger, A. oryzae, Rhizopus species,
Trichoderma species, and Fusarium culmorum was only slightly influenced
[79]. On the other hand, patulin inhibited in a modified filter disk assay the

sporulation of many common molds, especially of the Phycomycetes Rhizopus nigricans and Thamnidium elegans, and caused a reduction of spore germination in Cladosporium herbarum, Rhizopus nigricans, and Thamnidium elegans [80].

Treatment of Saccharomyces cerevisiae with patulin in concentrations up to 100 μg/ml did not affect fermentative CO_2 production [29]. Exposure of S. cerevisiae cells to 50 μg patulin/ml for 3 h during the exponential phase resulted in the formation of respiration-deficient mutants in a frequency of 60 to 80% [81].

4.3.3   Effects on Bacteria

It was demonstrated by many studies that patulin is a potent antibiotic [79]. The toxicity of patulin is very probably due to an interaction of the unsaturated keto group with the sulfhydryl group of enzymes vital for bacterial metabolism [79, 82]. This is confirmed by the observation that additions of SH compounds (e.g., cysteine) reduced the bacteriostatic property of patulin [79]. The sensitivity of Bacillus subtilis led to the development of a bioassay procedure by which as little as 1 μg toxin may be detected in a modified paper disk assay procedure [83].

4.4   Rubratoxin B

4.4.1   Effects on Higher Plants

Rubratoxin B (Sec. 2.3.4) acted as a strong inhibitor of the germination of Lepidium sativum seeds where the development of the hypocotyl was influenced more strongly than the development of the radicula. Rubratoxin B proved to be more inhibitory than aflatoxin B_1 [3] (Table 1).

A concentration of 0.1 μg rubratoxin B/ml acted as IAA synergist, whereas concentrations of more than 1 μg/ml were antagonistic to IAA [71] (Table 2).

The nuclei of root meristem cells of Allium cepa which had been treated with different concentrations of rubratoxin B showed indistinct nucleoli and anaphase bridges; the mitotic frequency was reduced [73].

4.4.2   Effects on Algae

Rubratoxin B did not influence the growth of Chlorella pyrenoidosa [22] but inhibited Volvox aureus [84].

4.4.3   Effects on Fungi

Concentrations of 100 μg/disk inhibited the sporulation of several fungi. In addition severe damage was observed in vegetative hyphae of Aspergillus niger, A. flavus, and Rhizopus nigricans after growth in rubratoxin B-containing medium. The hyphal tips of the Aspergillus species were often ruptured and

Fig. 4   Damage in hyphae of <u>Aspergillus</u> <u>niger</u> after cultivation for 4 days in the presence of 100 μg rubratoxin B/ml.   (a) Plasmoptysis, (b) giant cells.

the cytoplasm extruded (Fig. 4a), or the hyphae were swollen and greatly
septate, or were transformed into giant cells (Fig. 4b) whereas the germ tubes
of R. nigricans were much deformed [85]. It is assumed that rubratoxin B
interferes with cell wall synthesis. Concentrations up to 10 μg/ml enhanced
the fermentation activity of Saccharoymyces cerevisiae [29].

4.4.4   Effects on Bacteria

Rubratoxin B did not influence the growth of Bacillus megaterium [22], B.
subtilis [83], and other species of Bacillus, Micrococcus, and Staphylococcus
[84].

4.5   Byssochlamic Acid

Byssochlamic acid is produced by Paecilomyces varioti (see Sec. 1.1.2 for
key to genus) which is an important spoilage fungus in fruit juices.

4.5.1   Effects on Higher Plants

The germination of seeds of Sinapis alba was inhibited by byssochlamic acid and
its sodium salt in concentrations of more than 1:500 (the radicula was more
strongly influenced than the hypocotyl); the synthesis of chlorophyll was not
influenced [86, 87].

4.5.2   Effects on Fungi and Bacteria

The toxin inhibited the fermentation activity of Saccharomyces cerevisiae but
had no effects on E. coli [87].

4.6   Diacetoxyscirpenol

4.6.1   Effects on Higher Plants

Brian et al. [88] isolated a compound from culture filtrates of Fusarium
equiseti (Sec. 1.4.6.7) which they called "diacetoxyscirpenol" (Sec. 2.5.1.3.7).
This substance in lower concentrations (less than 2.5 μg/ml) enhanced the
development of roots of cress seedlings and was inhibitory at more than 10
μg/ml. The formation of the hypocotyl of cress seedlings was also inhibited
[71] (Table 1). The inhibitory action of diacetoxyscirpenol on IAA was pronounced [71] (Table 2).

Reduction of the frequency of mitoses and chromosomal aberrations was the result of a treatment of Allium cepa root tips with the toxin [73].

4.6.2   Effects on Algae

The growth of Chlorella pyrenoidosa was markedly suppressed by diacetoxy-scirpenol [22].

4.6.3   Effects on Fungi

Treatment of spores of different fungi with the toxin resulted in a significant inhibition of germ-tube extension [88] and sporulation [80]. The fermentation activity of bakers' yeast was enhanced by 0.001 to 1 μg toxin/ml and inhibited by 10 and 100 μg/ml [29].

4.6.4   Effects on Bacteria

Bacteria seem to be not very sensitive toward diacetoxyscirpenol. Only higher concentrations were inhibitory to Micrococcus pyogenes var. aureus, E. coli, and Bacillus subtilis [88, 88a] and 1 mg/ml in a disk assay procedure had only little effect on B. megaterium [46].

4.7   Other Fusarium Mycotoxins

4.7.1   Zearalenone F-2 Toxin

This toxin is produced by Fusarium roseum "graminearum" (Sec. 1.4.6.12) and acts as an estrogen toward fungi. Treatment of Cochliobolus carbonum and C. heterostrophus with 50 to 200 μg toxin resulted in a reduction of the number of perithecia; lower concentrations resulted in an increased number of these organs [89]. In F. roseum perithecial formation was enhanced as much as 100% after treatment with 0.1 to 10 μg [90]. Amounts in excess of 10 μg inhibited perithecia formation.

Only higher concentrations (1000 μg/ml) inhibited the growth of Chlorella pyrenoidosa [22] and Bacillus megaterium [46].

4.7.2   T-2 Toxin

As little as 4 μg/ml of the toxin (Sec. 2.5.1.3) (produced by F. tricinctum, Sec. 1.4.6.3) inhibited the germination of pea seeds totally [91]. After immersion of pea seedlings into aqueous solutions of T-2 toxin Marasas et al. [92] observed severe wilting, necrosis, and reduction in the average fresh weight and length of the seedlings dependent on the concentration. Symptoms were detectable with as little as 2.5 ppm toxin. When hypocotyl stems of soybean (<u>Glycine</u> <u>max</u>) were incubated in the presence of T-2 the elongation was inhibited as well as the growth-promoting effect of the auxin 2,4-D [93]. The properties of the plasma membrane, as measured by electrolyte leakage, were not affected by the toxin.

T-2 toxin acted fungistatic on <u>Rhodotorula</u> <u>rubra</u>, <u>R</u>. <u>glutinis</u>, <u>Saccharo</u>-<u>myces</u> <u>carlsbergensis</u>, <u>S</u>. <u>pastorianus</u>, <u>Penicillium</u> <u>digitatum</u>, and <u>Mucor</u> <u>ramannianus</u> (50 μg/assay disk) but was not bacteriostatic against 54 strains [91].

4.7.3   Toxic Butenolide

Even 200 μg/ml of this toxin (Sec. 2.5.1.3) (produced by <u>F</u>. <u>tricinctum</u>) was inactive toward the germination of pea seeds [91]. Different bacteria were weakly inhibited by 200 μg toxin/assay disk [91].

4.7.4   Moniliformin

A toxicogenic strain of <u>Fusarium</u> <u>moniliforme</u> (Sec. 1.4.6.6) was isolated in the United States from southern leaf blight-damaged corn seed [94]. Moni-liformin (Sec. 2.5.1.3), in a concentration of 20 ppm, inhibited the growth of wheat coleoptiles at 24%. When toxin solutions were sprayed onto tobacco seedlings, necrosis and interveinal chlorosis, distortion of leaf shape, thickening of the midrib, and, as a later effect, rosetting (probably caused by destruction of apical dominance) were the most striking symptoms of intoxica-tion.

4.8   Cytochalasins

4.8.1   Effects on Higher Plants

Cytochalasin B (Sec. 2.9.2), an effective inhibitor of intracellular streaming processes involving microfilament structures, inhibited the cytoplasmic movement in pollen tubes of <u>Tradescantia</u> <u>paludosa</u> in direct correlation to the concentration of the toxin [95]. Similar effects were observed in pollen tubes of <u>Lilium</u> <u>longiflorum</u>, rhizoids of the marine green alga <u>Caulerpa</u> <u>prolifera</u>, in <u>Acetabularia</u> <u>mediterranea</u>, and in the root hairs of <u>Lepidium</u> <u>sativum</u> and <u>Raphanus</u> <u>sativus</u> [96] as well as in cells of <u>Nitella</u> species and in <u>Avena</u> <u>sativa</u> seedlings [97]. The toxin did not cause ultrastructural changes in <u>Nitella</u> microfilaments as it does in some animal cell filaments.

Necrosis was observed in chicory, potato, and <u>Vinca</u> <u>minor</u> under the influence of cytochalasin B [98]. The inhibition of the growth of root hairs of <u>Lactuca</u> <u>sativa</u> was partially reversed by benzyladenine and kinetin [99].

4.8.2   Effects on Fungi and Bacteria

Cytochalasin A (Sec. 2.9.2) inhibited the growth of <u>Bacillus subtilis</u>, <u>E. coli</u>,
and different fungi [100]. The cytochalasins A and D showed antifungal properties. In <u>Botrytis cinerea</u> enhanced branching and swelling of hyphal tips were
significant. Cytochalasin B was neither antibacterial nor antifungal [101]. In
the course of spore germination of <u>Botrytis cinerea</u> the germination hyphae
were swollen and distorted and more strongly branched than controls [102].
Oliver [103] observed frequently branching and beaded filaments in <u>Asper-
gillus nidulans</u> under the influence of cytochalasin B. In <u>Polyporus biennis</u>
the toxin inhibited the radial growth rate and caused an increase in hyphal
density through a reduction in the distance between successive branches. The
position of clamp connections was only slightly affected [104].

4.9   Other Mycotoxins

Penicillic acid (Sec. 2.3.6) [77, 79] and aspergillic acid [77] showed certain
antibacterial and antifungal properties. The toxicity of penicillic acid was
suggested to be due to an interaction with sulfhydryl compounds [79]. Citrinin
inhibited the germination of seeds of white mustard and red clover [105].
Sporidesmin suppressed the germination of cress seeds in a concentration of
45 μg/ml [106]. The growth of <u>Bacillus cereus</u> mycoides was suppressed in a
disk assay technique by as little as 1.5 μg ochratoxins [107]. Oosporein from
<u>Chaetomium trilaterale</u> reduced the growth of <u>Avena sativa</u> internode sections
and inhibited the growth of coleoptiles. Treated tobacco plants showed stunting,
rosetting, leaf puckering, and light chlorosis [108].

4.10   Recent Developments

4.10.1   Aflatoxins

The treatment of roots of peanut seedlings with 10 to 100 ppm aflatoxin B_1 was
inhibitory to the elongation and dry weight of roots and stems [109]. A concentration of 2 μg aflatoxin B_1/ml reduced the germination of pollen grains of
<u>Lilium longiflorum</u> and the elongation of pollen tubes was gradually suppressed
by toxin levels of more than 6 μg/ml [110]. In <u>Kalanchoe daigremontiana</u> the
elongation of roots was inhibited by 100 μg aflatoxin B_1/ml [111].
    In their studies on the degradation of aflatoxin B_1 by microorganisms Mann
and Rehm [112] found that <u>Corynebacterium rubrum</u> degraded more than 99% of
added toxin (1.48 μg/ml) after 4 days and that the anascosporogenic yeast
(<u>Candida</u> spp.) degraded at least 80% of aflatoxin B_1 within 20 days. Molds of
the genera <u>Aspergillus</u> and <u>Penicillium</u> removed the toxin to a considerable
extent. It is suggested that aflatoxin B_1 is degraded to aflatoxin R_0 by various
molds.

When reevaluating the sensitivity of Bacillus megaterium toward aflatoxin
B_1 in a disk assay procedure Jamicki et al. [113] observed zones of inhibition
at 0.13 μg/disk and a linear dependence between the diameter of the zone of
inhibition and the amount of toxin in the disk within the range of 0.26 to 1.25
μg/disk.

Ames et al. [114] used the sensitivity of a histidine mutant of Salmonella
typhimurium to detect the mutagenic ability of a compound that has been produced by activating aflatoxin B_1 by treatment with rat liver homogenates. In
similar experiments, Schoenhard et al. [115] observed that a DNA repair-
deficient strain of Bacillus subtilis, insensitive to high levels of aflatoxin B_1,
was reduced in viability after incubation in the supernatant of liver homogenates
of rainbow trouts (Salmo gairdneri) which had been preincubated with aflatoxin
B_1.

4.10.2   Patulin

Extensive work on the influence of a single dose of patulin on the growth of
wheat revealed that this plant is unusually susceptible to patulin at two significant growth periods: at germination and during stem elongation and the
heading and flowering period [116]. The development of roots of epiphyllous
buds of Kalanchoe daigremontiana was inhibited by 100 μg patulin/ml [111].

Strains of the yeasts Saccharomyces cerevisiae and S. ellipsoideus were
able to degrade patulin in apple juice [117]. Stott and Bullerman [118]
developed a filter disk biotest for patulin based on the sensitivity of Bacillus
megaterium NRRL 1368. The response of this organism to the toxin was linear
between 2 and 80 μg. Bacillus subtilis, Mycobacterium smegmatis, and Photobacterium fischeri were most sensitive toward patulin [119]. Treatment of
Bacillus subtilis cells with 100 μg patulin/disk inhibited the activity of cytochrome oxidase as well as the citrate utilization and the mannitol fermentation,
and caused the development of elongated cells [120].

4.10.3   Rubratoxin B

Treatment of buds of Kalanchoe daigremontiana with 100 and 10 μg rubratoxin
B/ml totally inhibited the development of roots and reduced the length of the
first internode [111].

4.10.4   Citrinin

In their study on the influence of various antibiotics on auxin activity in higher
plants Iyengar and Starkey [112] found that citrinin reduced the growth-inducing
effect of β-indolylacetic acid. A treatment of epiphyllous buds of Kalanchoe
daigremontiana with citrinin strongly reduced the formation of roots [111].
Wilting and leaf damage in beans, cotton, and sorghum by citrinin were
reported by Indian authors [122].

Citrinin was supposed to be a powerful antibiotic because of its bacterio-static properties against staphylococci and other bacteria in dilutions of 1:15,000 to 1:50,000 [119, 123, 124]. Treatment of Staphylococcus aureus with citrinin resulted in an inhibition of respiratory enzymes [125].

The phytopathogenic fungi Sclerotium rolfsii and Endothia parasitica were inhibited by pure citrinin and by the citrinin-forming mold Penicillium citrinum [126].

4.10.5   Penicillic Acid

In a screening test penicillin acid proved to be more or less inhibitory to various bacteria (including B. enteritides, S. paratyphi, E. coli, S. aureus) [119, 123, 124].

4.10.6   Ochratoxin A

Only gram-positive bacteria were inhibited by ochratoxin A, generally at a pH lower than 7.0. During growth inhibition protein and RNA synthesis were reduced simultaneously [127].

4.10.7   Sterigmatocystin

Lillehoj and Ciegler [128] observed no sensitivity of various bacteria, yeasts (S. pastorianus, Candida albicans, Torulopsis utilis), and molds (P. brevi-compactum, P. urticae, P. cyclopium, P. citreo-viride, A. versicolor, A. flavus) toward sterigmatocystin in a disk assay test with 100 μg toxin/disk.

4.10.8   Fusarium Mycotoxins

When seedlings of various field crops were dipped into the culture broth of 642 Fusarium isolates, a correlation between the degree of phytotoxicity and the grade of animal toxicity was observed [129].

4.10.9   Luteoskyrin

A concentration of 0.4 μg luteoskyrin/ml caused the 50% inhibition of the growth of E. coli F-11 [130].

Acknowledgements

I would like to express my gratitude to Professor E. V. Crisan (Department of Food Science and Technology, University of California, Davis, California) and Professor F. Grossmann (Lehrstuhl für Phytopathologie und Pflanzenschutz, Universität Hohenheim, Germany) for their donation of photographs.

References

1. A. Graniti (1972): In R. K. S. Wood, A. Ballio, and A. Graniti (eds.): Phytotoxins in Plant Diseases. Academic Press, New York, pp. 1-18.
2. R. Schoental and A. F. White (1965): Aflatoxins and albinism in plants. Nature (London) 205:57.
3. J. Reiss (1971): Hemmung der Keimung der Kresse (Lepidium satiuum) durch Aflatoxin B_1 and Rubratoxin B. (Inhibition of germination of Lepidium sativum by aflatoxin B_1 and rubratoxin B.) Biochem. Physiol. Pflanzen 162:363.
4. E. V. Crisan (1973): Effects of aflatoxin on germination and growth of lettuce. Appl. Microbiol. 25:342.
5. E. V. Crisan (1973): Effects of aflatoxin on seedling growth and ultrastructure in plants. Appl. Microbiol. 26:991.
5a. A. A. Adekunle and O. Bassir (1973): The effects of aflatoxin B_1 and palmotoxins B_0 and G_0 on the germination and leaf colour of the cowpea (Vigna sinensis), Mycopathol. Mycol. Appl. 51:299.
6. J. Reiss (1971): Föderung der Aktivität von β-Indolylessigsäure durch Aflatoxin B_1. Z. Pflanzenphysiol. 64:260.
7. H. S. Black and A. M. Altschul (1965): Gibberellic acid-induced lipase and alpha-amylase formation and their inhibition by aflatoxin. Biochem. Biophys. Res. Commun. 19:661.
8. H. C. Jones, H. S. Black, and A. M. Altschul (1967): Comparison of effects of gibberellic acid and aflatoxin in germinating seeds. Nature (London) 214:171.
9. B. Truelove, D. E. Davis, and O. C. Thompson (1970): Effects of aflatoxin B_1 on protein synthesis by cucumber cotyledon discs. Can. J. Bot. 48:485.
10. T. Asahi, Z. Mori, R. Majima, and I. Uritani (1969): The effects of aflatoxins on metabolic changes in plant tissue in response to injury. J. Stored Prod. Res. 5:219.
11. I. Uritani, T. Asahi, R. Majima, and Z. Mori (1970): The biochemical effects of aflatoxins and other toxic compounds related to parasitic fungi on the metabolism of plant tissue, in M. Herzberg (ed.): Proceedings of the First U.S.-Japan Conference on Toxic Micro-Organisms. U. S. Dept. of the Interior, Washington, D.C., pp. 107-113.
11a. W. V. Dashek and G. C. Llewellyn (1974): The influence of the carcinogen aflatoxin B_1 on the metabolism of germinating lily pollen. In H. F. Linskens (ed): Fertilization in Higher Plants. North-Holland Publ., Amsterdam, pp. 351-360.
12. A. Z. Joffe (1969): Effects of Aspergillus flavus on groundnuts and on some other plants. Phytopathol. Z. 64:321.
13. J. Reiss (1969): Hemmung des Sprosswachstums von Caralluma frerei Rowl. durch Aflatoxin. Planta 89:369.
14. M. El-Khadem, G. Menke, and F. Grossmann (1966): Schädigung von Erdnusskeimlingen durch Aflatoxine. Naturwissenschaften 53:532.
15. M. El-Khadem (1968): Die Bedeutung von Aflatoxin für die durch Aspergillus flavus verursachte Keimlingskrankheit der Erdnuss. Phytopathol. Z. 61:218.

16. L. J. Lilly (1965): Induction of chromosome aberrations by aflatoxin. Nature (London) 207:433.

17. J. Jacquet and P. Boutibonnes (1970): Effets des flavacoumarines (aflatoxines) sur quelques animaux et végétaux. C.R. Soc. Biol. 164:2239.

18. J. Reiss (1971): Chromosomenaberrationen in den Wurzelspitzen von Allium cepa durch Aflatoxin B₁. Experientia 27:971.

18a. J. Jacquet, P. Boutibonnes, and S. Saint (1971): Effets biologiques des flavacoumarines d'Aspergillus parasiticus ATCC 15517. T. Animaux. Rev. Immunol. 35:159.

19. I. Slowatizky, A. M. Mayer, and A. Poljakoff-Mayber (1969): The effect of aflatoxin on greening of etiolated leaves. Israel J. Bot. 18:31.

20. A. M. Mayer, A. Poljakoff-Mayber, P. Robinson, and I. Slowatizky (1969): A simple bioassay for detection of aflatoxin in milk. Toxicon 7:13.

21. M. Ikawa, D. S. Ma, G. M. Meeker, and R. P. Davis (1969): Use of Chlorella in mycotoxin and phytotoxin research. J. Agric. Food Chem. 17:425.

22. J. D. Sullivan, Jr. and M. Ikawa (1972): Variations in inhibition of growth of five Chlorella strains by mycotoxins and other toxic substances. J. Agric. Food Chem. 20:921.

23. E. B. Lillehoj, A. Ciegler, and H. H. Hall (1967): Fungistatic action of aflatoxin B₁. Experientia 23:187.

24. J. Reiss (1970): Untersuchungen über den Einfluss von Aflatoxin B₁ auf die Morphologie und die cytochemisch fassbare Aktivität einiger Enzyme von Mucor hiemalis (Mucorales). Mycopathol. Mycol. Appl. 42:225.

25. J. Reiss (1971): Inhibition of fungal sporulation by aflatoxin. Arch. Mikrobiol. 76:219.

26. J. Reiss (1971): Hyphenanomalien bei Thamnidium elegans Link durch Aflatoxin B₁. Z. Allg. Mikrobiol. 11:637.

27. J. D. Wildman (1966): Note on occurrence of giant cells in Aspergillus flavus Link. J. Assoc. Off. Anal. Chem. 49:562.

28. J. Jacquet and P. Boutibonnes (1969): Actions de l'aflatoxine sur les cellules microbiennes. C.R. Soc. Biol. 163:2574.

29. J. Reiss (1973): Influence of the mycotoxins aflatoxin B₁, rubratoxin B, patulin and diacetoxyscirpenol on the fermentation activity of baker's yeast. Mycopathol. Mycol. Appl. 51:337.

30. T. Ong (1970): Mutagenicity of aflatoxins in Neurospora crassa. Mutation Res. 9:615.

31. T. Ong (1970): Mutagenicity and mutagenic specificity of aflatoxin B₁ and G₁ in Neurospora crassa. Genetics 64:s48.

32. T. Ong (1971): Mutagenic activities of aflatoxin B₁ and G₁ in Neurospora crassa. Mol. Gen. Genetics 111:159.

33. T. Ong and F. J. de Serres (1972): Mutagenicity of chemical carcinogens in Neurospora crassa. Cancer Res. 32:1890.

33a. R. W. Detroy and C. W. Hesseltine (1969): Transformation of aflatoxin B₁ by steroid-hydroxylating fungi. Can. J. Microbiol. 15:495.

34. R. J. Cole and J. W. Kirksey (1971): Aflatoxin G₁ metabolism by Rhizopus species. J. Agric. Food Chem. 19:222.

35. R. J. Cole, J. W. Kirksey, and B. R. Blankenship (1972): Conversion of aflatoxin B_1 to isomeric hydroxy compounds by Rhizopus spp. J. Agric. Food Chem. 20:1100.

36. H. R. Burmeister and C. W. Hesseltine (1966): Survey of sensitivity of microorganisms to aflatoxin. Appl. Microbiol. 14:403.

37. E. B. Lillehoj and A. Ciegler (1968): Aflatoxin B_1 binding and toxic effects on Bacillus megaterium. J. Gen. Microbiol. 54:185.

38. L. R. Beuchat and R. V. Lechowich (1971): Morphological alterations in Bacillus megaterium as produced by aflatoxin B_1. Appl. Microbiol. 21:124.

39. L. R. Beuchat and R. V. Lechowich (1971): Biochemical alterations in Bacillus megaterium as produced by aflatoxin B_1. Appl. Microbiol. 21:119.

40. H. R. Burmeister and C. W. Hesseltine (1967): Aflatoxin sensitivities of an L form and Bacillus megaterium. Bacteriol. Proc.: 17.

41. E. B. Lillehoj and A. Ciegler (1970): Aflatoxin B_1 induction of lysogenic bacteria. Appl. Microbiol. 20:782.

42. N. L. Clements (1968): Note on a microbiological assay for aflatoxin B_1: A rapid confirmatory test by effects on growth of Bacillus megaterium. J. Assoc. Off. Anal. Chem. 51:611.

43. N. L. Clements (1968): Rapid confirmatory test for aflatoxin B_1, using Bacillus megaterium. J. Assoc. Off. Anal. Chem. 51:1192.

44. A. Jayaraman, E. J. Herbst, and M. Jkawa (1968): The bioassay of aflatoxins and related substances with Bacillus megaterium spores and chick embryos. J. Am. Oil Chem. Soc. 45:700.

45. L. Viitasalo and H. G. Gyllenberg (1968): Toxicity of aflatoxins to Bacillus megaterium. Lebensmittel-Wiss. u. Technol., 2:113

46. A. R. Buckelew, Jr., A. Chakravarti, W. R. Burge, V. M. Thomas, Jr., and M. Ikawa (1972): Effect of mycotoxins and coumarins on the growth of Bacillus megaterium from spores. J. Agric. Food Chem. 20:431.

47. E. B. Lillehoj, A. Ciegler, and H. H. Hall (1967): Aflatoxin B_1 uptake by Flavobacterium aurantiacum and resulting toxic effects. J. Bacteriol. 93:464.

48. E. B. Lillehoj and A. Ciegler (1967): Inhibition of deoxyribonucleic acid synthesis in Flavobacterium aurantiacum by aflatoxin B_1. J. Bacteriol. 94:787.

49. E. B. Lillehoj, A. Ciegler, and H. H. Hall (1967): Aflatoxin G_1 uptake by cells of Flavobacterium aurantiacum. Can. J. Microbiol. 13:629.

50. E. B. Lillehoj, R. D. Stubblefield, G. M. Shannon, and O. L. Shotwell (1971): Aflatoxin M_1 removal from aqueous solutions by Flavobacterium aurantiacum. Mycopathol. Mycol. Appl. 45:259.

51. J. B. Wragg, V. C. Ross, and M. S. Legator (1967): Effect of aflatoxin B_1 on deoxyribonucleic acid polymerase of Escherichia coli. Proc. Soc. Exp. Biol. Med. 31:1052.

52. O. U. Eka (1972): Effect of aflatoxins on microorganisms. Z. Allg. Mikrobiol. 12:593.

53. P. Boutibonnes and J. Jacquet (1972): Recherches complémentaires sur les effets biologiques des flavacoumarines. Rev. Immunol. 36:85.

54. J. Jacquet, P. Boutibonnes, and S. Saint (1972): Effets biologiques des flavacoumarines sur les bactéries mise en évidence par des méthodes de diffusion en gélose. Bull. Acad. Vét. 45:147.

55. P. Boutibonnes (1969): Recherches sur les principaux Aspergillus pathogènes. Rev. Immunol. 33:177.

56. J. Reiss (1975): Mycotoxin bioassay, using Bacillus stearothermophilus. J. Assoc. Off. Anal. Chem. 58:624.

57. P. Boutibonnes (1972): (French) Biological effects of aflatoxins of Aspergillus parasiticus Attc 15 517 3. Microorganisms. Rev. Immunol. 36:15.

58. T. Arai, T. Ito, and Y. Koyama (1967): Antimicrobial activity of aflatoxins. J. Bacteriol. 93:59.

59. J. Nezval, H. Bösenberg, and U. Linzel (1970): Untersuchungen über die Aflatoxinwirkung auf Bakterien. Arch. Hyg. 154:143.

60. E. B. Lillehoj and A. Ciegler (1970): Aflatoxin B_1 effect on enzyme biosynthesis in Bacillus cereus and Bacillus licheniformis. Can. J. Microbiol. 16:1059.

61. T. M. McCalla, W. D. Guenzi, and F. A. Norstadt (1963): Microbial studies of phytotoxic substances in the stubble-mulch system. Z. Allg. Mikrobiol. 3:202.

62. F. A. Norstadt and T. M. McCalla (1963): Phytotoxic substance from a species of Penicillium. Science 140:410.

63. T. M. McCalla and F. A. Norstadt (1967): Influence of patulin on the growth of wheat plants. Bacteriol. Proc. A98.

64. F. A. Norstadt and T. M. McCalla (1971): Effects of patulin on wheat grown to maturity. Soil Sci. 111:236.

65. J. R. Ellis and T. M. McCalla (1973): Effects of patulin and method of application on growth stages of wheat. Appl. Microbiol. 25:562.

66. M. I. Timonin (1946): Activity of patulin against Ustilago tritici (Pers.) Jen. Sci. Agric. 26:358.

67. H. Börner (1963): Untersuchungen über die Bildung antiphytotischer und antimikrobieller Substanzen durch Mikrooganismen im Boden und ihre mögliche Bedeutung für die Bodenmüdigkeit beim Apfel (Pirus malus L.) I. Bildung von Patulin und einer phenolischen Verbindung durch Penicillium expansum auf Wurzel- und Blattrückständen des Apfels. Phytopathol. Z. 48:370.

68. H. Börner (1963): (II. Untersuchungen...). Der Einfluss verschiedener Faktoren auf die Bildung von Patulin und einer phenolischen Verbindung durch Penicillium expansum auf Blatt- und Wurzelrückständen des Apfels. Phytopathol. Z. 49:1.

69. G. Miescher (1950): Über die Wirkungsweise von Patulin auf höhere Pflanzen, insbesondere auf Solanum lycopersicum L. Phytopathol. Z. 16:379.

69a. E. Gäumann and O. Jaag (1947): Die physiologischen Grundlagen des parasitogenen Welkens II. Ber. Schweiz. Bot. Ges. 57:132.

70. H. W. Klemmer, A. J. Riker, and O. N. Allen (1955): Inhibition of crown gall by selected antibiotics. Phytopathology 45:618.

70a. V. R. Wallen and A. J. Skolko (1951): Activity of antibiotics against Ascochyta pisi. Can. J. Bot. 29:316.

71. J. Reiss (1973): Untersuchungen über den Einfluss von Mycotoxinen auf die Aktivität der β-Indolylessigäure und die Keimung von Kressesamen (Lepidium sativum). Z. Pflanzenphysiol. 69:274.

72. E. Steinegger and H. Leupi (1956): Untersuchungen über den Einfluss von Pflanzenwirkstoffen auf das Wurzelwachstum von Allium cepa und die Keimung von Lepidium sativum. Pharm. Acta Helv. 31:45

73. J. Reiss (1975): Mycotoxin poisoning of Allium cepa root tips II. Reduction of mitotic index and formation of chromosomal aberrations and cytological abnormalities by patulin, rubratoxin B and diacetoxyscirpenol. Cytologia 40:703.

74. F. H. Wang (1948): The effects of clavacin upon root growth. Bot. Bull. Acad. Sinica 2:265.

75. C. C. Barnum (1924): The production of substances toxic to plants by Penicillium expansum Link. Phytopathology 14:238.

76. J. Reiss (1970): Untersuchungen über die Phytotoxizität von Penicillium expansum, Aspergillus niger, und Rhizopus nigricans. Phytopathol. Z. 69:78.

77. K. Gilliver (1946): The inhibitory action of antibiotics on plant pathogenic bacteria and fungi. Ann. Bot. 10:271.

78. W. K. Anslow, H. Raistrick, and G. Smith (1943): Anti-fungal substances from moulds. Part I. Patulin (anhydro-3-hydroxymethylene-tetrahydro-1,4-pyrone-2-carboxylic acid), a metabolic product of Penicillium patulum Banier and Penicillium expansum (Link) Thom. J. Soc. Chem. Ind. 62:236.

79. B. Geiger and J. E. Conn (1945): The mechanism of the antibiotic action of clavacin and penicillic acid. J. Am. Chem. Soc. 67:112.

80. J. Reiss (1973): Influence of the mycotoxins patulin and diacetyscirpenol on fungi. J. Gen. Appl. Microbiol. 19:415.

81. V. M. Mayer and M. S. Legator (1969): Production of petite mutants of Saccharomyces cerevisiae by patulin. J. Agric. Food Chem. 17:454.

82. M. Rinderknecht, J. L. Ward, F. Bergel, and A. L. Morrison (1947): Studies on antibiotics. 2. Bacteriological activity and possible mode of action of certain non-nitrogenous natural and synthetic antibiotics. Biochem. J. 41:463.

83. J. Reiss (1975): Bacillus subtilis; a sensitive bioassay for patulin. Bull. Environm. Contam. Toxicol. 13:689.

84. A. W. Hayes and E. P. Wyatt (1970): Survey of sensitivity of micro-organisms to rubratoxin B. Appl. Microbiol. 29:164.

85. J. Reiss (1972): Toxicity of rubratoxin B to fungi. J. Gen. Microbiol. 71:167.

86. H. Meyer and H.-J. Rehm (1969): Keimungs-und garungshemmende Wirkung von Byssochlaminsäure. Naturwissenschaften 55:563

87. H. Meyer (1971) Ph.D. Thesis: Wilhelms-Universität, Münster/Westf.

88. P. W. Brian, A. W. Dawkins, J. F. Grove, H. G. Hemming, D. Lowe, and G. L. F. Norris (1961): Phytotoxic compounds produced by Fusarium equiseti. J. Exp. Bot. 12:1.

88a. H. Stähelin, M. E. Kalberer-Rüsch, E. Signer, and S. Lazáry (1968): Über einige biologische Wirkungen des Cytostaticum Diacetoxyscirpenol. Arzneim. Forsch. 18:989.

89. R. R. Nelson, C. J. Mirocha, D. Huisingh, and A. Tijerina-Menchaca (1968): Effects of F_2, an estrogenic metabolite from Fusarium, on sexual reproduction of certain ascomycetes. Phytopathology 58:1061.

90. J. C. Wolf and C. J. Mirocha (1973): Regulation of sexual reproduction in Gibberella zeae (Fusarium roseum graminearum) by F_2 (zearelenone). Can. J. Microbiol. 19:725.

91. H. R. Burmeister and C. W. Hesseltine (1970): Biological assays for two mycotoxins produced by Fusarium tricinctum. Appl. Microbiol. 20:437.

92. W. F. O. Marasas, E. B. Smalley, J. R. Bamburg, and F. M. Strong (1971): Phytotoxicity of T-2 toxin produced by Fusarium tricinotum. Phytopathology 61:1488.

93. C. Stahl, L. N. Vanderhoef, N. Siegel, and J. P. Helgeson (1973): Fusarium tricinctum T$_2$ toxin inhibits auxin-promoted elongation in soybean hypocotyl. Plant Physiol. 52:663.

94. R. J. Cole, J. W. Kirksey, H. G. Cutler, B. L. Doupnik, and J. L. Peckham (1973): Toxin from Fusarium moniliforme: Effects on plants and animals. Science 179:1324.

95. J. P. Mascarenhas and J. Lafountain (1972): Protoplasmic streaming, cytochalasin B, and growth of the pollen tube. Tissue Cell 4:11.

96. W. Herth, W. W. Franke, and W. J. Vanderwoude (1972): Cytochalasin stops tip growth in plants. Naturwissenschaften 59:38.

97. M. O. Bradley (1973): Microfilaments and cytoplasmic streaming: Inhibition of streaming with cytochalasin. J. Cell Sci. 12:327.

98. J. F. Bousquet and M. Barbier (1972): Sur l'activité phytotoxique de trois souches de Phoma exigua et la présence de la cytochalasine B (ou phomine) dans leur milieu de culture. Phytopathol. Z. 75:365.

99. V. K. Sawhney and L. M. Srivastava (1974): Cytochalasin B-induced inhibition of root-hair growth in lettuce seedlings and its reversal by benzyladenine. Planta 119:165.

100. V. Betina and D. Mičeková (1972): Antimicrobial properties of fungal macrolide antibiotics. Z. Allg. Mikrobiol. 12:355.

101. V. Betina, D. Mičeková, and P. Nemec (1972): Antimicrobial properties of cytochalasins and their alteration of fungal morphology. J. Gen. Microbiol. 71:343.

102. V. Betina and D. Mičeková (1973): Morphogenetic activity of cytochalasins, cyanein and monorden in Botrytis cinerea. Z. Allg. Mikrobiol. 13:287.

103. P. T. P. Oliver (1973): Influence of cytochalasin B on hyphal morphogenesis of Aspergillus nidulans. Protoplasma 76:279.

104. A. M. Patton and R. Marchant (1975): Effect of cytochalasin B on hyphal morphogenesis in Polyporus biennis. J. Gen. Microbiol. 86:301.

105. J. M. Wright (1951): Phytotoxic effects of some antibiotics. Ann. Bot. 15:493.

106. D. E. Wright (1968): Toxins produced by fungi. Annu. Rev. Microbiol. 22:269.

107. D. Broce, R. M. Grodner, R. L. Killebrew, and F. L. Bonner (1971): Ochratoxins A and B confirmation by microbiological assay using Bacillus cereus mycoides. J. Assoc. Off. Anal. Chem. 53:616.

108. R. J. Cole, J. W. Kirksey, H. G. Cutler, and E. E. Davis (1974): Toxic effects of oosporein from Chaetomium trilaterale. J. Agric. Food Chem. 22:517.

109. M. El-Khadem, M. S. Tewfik, and Y. A. Hamdi (1975): Effect of aflatoxin B$_1$, aspergillic and kojic acid on peanut seedlings. Zbl. Bakt. II 130:556.

110. G. C. Llewellyn and W. V. Dashek (1973): The influence of the hepatocarcinogen aflatoxin B$_1$ on lily pollen germination and tube elongation. Incompatability Newsletters, No. 3, pp. 18-22.

111. J. Reiss (1977): Effects of mycotoxins on the development of epiphyllous buds of Kalandroe daigremontiana. Z. Pflanzenphysiol. 82:446.

112. R. Mann and H. J. Rehm (1975): Degradation of aflatoxin B₁ by micro-organisms. Naturwissenschaften 62:537.

113. J. Jamicki, S. Stawicki, K. Szebiotko, B. Cozaś, M. Kokorniak, and M. Wiewiórowska, The utilization of the microbiological test for determining the bacteriostatic activity of the fluorescent substances produced by the microflora of wheat grain. Acta Aliment. Polon. 25:107.

114. B. N. Ames, W. E. Durston, E. Yamasaki, and F. D. Lee (1973): Carcin-ogens are mutagens: A simple text system combining liver homogenates for activation and bacteria for detection. Natl. Acad. Sci. U.S.A. 70:2281.

115. G. L. Schoenhard, P. E. Bishop, D. J. Lee, and R. O. Sinnhuber (1975): Bacillus subtilis GSY 1057 assay for aflatoxin B₁ activation by rainbow trout (Salmo gairdneri). J. Assoc. Off. Anal. Chem. 58:1074.

116. T. M. McCalla and F. A. Norstadt (1974): Toxicity problems in mulch tillage. Agric. Environm. 1:153.

117. J. Harwig, P. M. Scott, B. P. C. Kennedy, and Y. -K. Chen (1973): Disappearance of patulin from apple juice fermented by Saccharomyces spp. J. Can. Inst. Food Sci. Technol. 6:45.

118. W. T. Stott and L. B. Bullerman (1975): Microbiological assay of patulin, using Bacillus megaterium. J. Assoc. Off. Anal. Chem. 58:497.

119. F. Kavanagh (1947): Activities of twenty-two antibacterial substances against nine species of bacteria. J. Bacteriol. 54:761.

120. J. Reiss (1976): Biochemical effects in Bacillus subtilis after treatment with the mycotoxin patulin. Z. Allg. Mikrobiol. 16:229.

121. M. R. S. Iyengar and R. L. Starkey (1953): Synergism and antagonism of auxin by antibiotics. Science 118:357.

122. C. Damodaran, S. Kathirvel-Pandian, S. Seeni, R. Selvam, M. G. Ganesan, and S. Shanmugasundaram (1975): Citrinin a phytotoxin? Experientia 31:1415.

123. A. E. Oxford (1942): Anti-bacterial substances from moulds. Part III. Some observations on the bacteriostatic powers of the mould products citrinin and penicillic acid. Chem. Ind. 61:48.

124. N. G. Heatley and F. J. Philpot (1947): The routine examination of anti-biotics produced by moulds. J. Gen. Microbiol. 1:232.

125. M. Michaelis and F. S. Thatcher (1945): The action of citrinin on some respiratory enzymes of Staphylococcus aureus and Escherichia coli. Arch. Biochem. Biophys. 8:177.

126. O. Verona and P. Gambogi (1952): Una ricerca sull'azione della citrinina prodotta da Penicillium citrinum Thom su funghi fitopatogeni. Phyto-pathol. Z., 19:423.

127. K. Keller, C. Schulz, R. Löser, and R. Röschenthaler (1975): The inhibition of bacterial growth by ochratoxin A. Can. J. Microbiol. 21:972.

128. E. B. Lillehoj and A. Ciegler (1968): Biological activity of sterigmato-cystin. Mycopathol. Mycol. Appl. 35:373.

129. A. Z. Joffee and J. Palti (1974): Relations between harmful effects on plants and on animals of toxins produced by species of Fusarium. Myco-pathol. Mycol. Appl. 52:209.

130. Y. Yamakawa and Y. Ueno (1970): Microbial assay of hepatotoxic authra-quinones with Escheridia coli T-11. Chem. Pharm. Bull. 18:177.

Jürgen Reiss

PART 5

CONTAMINATION BY MYCOTOXINS: WHEN IT OCCURS AND HOW TO
PREVENT IT

5.1   Introduction

Where and when do mycotoxins get into our foods and feedstuffs? How can we prevent such contamination? These are the principal questions dealt with in this section.

Mycotoxins, by definition, are toxic substances produced by fungi growing in feed or food products. If we could prevent fungi from growing in these commodities, we would of course have no mycotoxin problems. Prevention is the most direct and satisfactory means of dealing with this and many other problems. In other words, we should treat the causes, not just the symptoms.

It is possible in some situations to prevent the production of a toxin without preventing fungal growth. The insecticide dichlorovos, for example, inhibits synthesis of aflatoxin [27]. However, it would seem more practical and more desirable to expend efforts on a control measure which produces a sound, mold-free commodity rather than one free of aflatoxin, but otherwise spoiled.

Depending on the commodity, mycotoxin contamination is either a field problem, a storage problem, or a combination of the two. Since mycotoxins are produced by fungi, they should be viewed broadly as a potential danger anywhere fungi grow on materials which are used as food or feed. Fungal contamination is necessary for production of mycotoxins, but toxicity is certainly not the inevitable result of all fungal invasion. Fungi are almost universally present on and in cereal grains, nuts, and nearly all other plant materials, but toxicity seems to be the exception rather than the rule.

5.2   Mycotoxins Produced in the Field

Some mycotoxins, such as those that occur on pasture grasses, are strictly field problems. The toxin that causes facial eczema of ruminants is a well-known example of this type [5]. Pithomyces chartarum, which produces the facial eczema toxin, requires specific weather conditions for development. Periods of near 100% relative humidity must accompany mild temperatures if the fungus is to flourish. Control of this kind of mycotoxin problem is initially a matter of identifying and avoiding the contaminated material. A more permanent solution would be to control the fungus through cultural practices, chemicals, or resistant plant varieties.

Ergotism is one of the oldest mycotoxicoses. Occasional heavy infections of wheat and rye have contributed to human suffering for centuries, but contamination in recent years has been less serious [12]. Current varieties of small grains are fairly resistant, but suitable environmental conditions may cause localized outbreaks. If and when hybrid wheats are widely grown, ergotism may return to prominence. The hybrids seem to be susceptible to the causal fungus, Claviceps purpurea [21].

147

Mycotoxins that develop only in the field are often very sporadic and do not lend themselves to permanent solutions. Some suspected mycotoxicoses have been very elusive to researchers who are studying their etiology, at least partly because of their sporadic nature. Fescue foot [36] and Bermuda grass tremors [8] are examples. These diseases appear to be caused by mycotoxins, but proof has not been established.

Field-produced mycotoxins are usually associated with unusual weather conditions. An extremely wet harvest season may allow common Fusarium species, for example, to invade corn or other cereal grains heavily, often making the grain toxic or unpalatable. A number of the Fusarium mycotoxicoses are reviewed in the treatise edited by Kadis et al. [16]. Alimentary toxic aleukia is perhaps the most dramatic of the field-produced mycotoxin problems, having been responsible for the poisoning of thousands of people who consumed cereal grains which overwintered in the fields [14]. Fortunately such calamities do not occur frequently, because we are usually powerless to prevent them. Because exposure of mature crops to wet weather is more likely to result in toxicity or other deterioration, every effort should be made to get crops out of the field and properly prepared for storage.

5.3   Aspergillus flavus Development
in the Field

Aspergillus flavus, along with other species of Aspergillus and Penicillium, have long been considered as problems only in stored products. Freshly harvested grains have often been shown to be essentially free of "storage fungi" [6, 32], and mycotoxin contamination by storage fungi was thought to be unlikely or at least insignificant in crops before normal harvest time. However, A. flavus often has been reported on corn before harvest. In 1920 Taubenhaus [31] published an extensive report on A. flavus and A. niger as ear mold problems in Texas. Other researchers have reported the incidence of at least small amounts of A. flavus in corn from the Midwest [17, 20, 33, 34]. A Russian report in 1962 indicated that in the Ukraine A. flavus was abundant in corn from the milk stage to maturity [25].

Following the initial discovery of aflatoxin, many commodities have been examined not just for A. flavus, but for aflatoxin content, both before and after harvest. The Quaker Oats Company began an intensive survey of corn fields in 1971 after preliminary findings indicated preharvest contamination (R. Wichser, personal communication). Since that time field occurrences of aflatoxin in corn have been observed and studied by others [18, 22]. These studies as well as surveys by private industry, the U.S. Department of Agriculture, and the Food and Drug Administration indicate that field contamination of corn in the United States is primarily a problem in the South and Southeast. Corn in marketing channels in those regions has been shown to have a higher incidence of aflatoxin than corn from the Midwest [28], but such studies cannot distinguish between preharvest and postharvest contamination.

Lillehoj et al. [18] reported that the incidence of aflatoxin was much higher in southeastern Missouri than in fields sampled in southern Illinois. Rambo et al. [22] reported a north-to-south increase in the incidence of A. flavus-infected kernels in Indiana, but did not find naturally occurring aflatoxin. They stated, "Field infections of corn by A. flavus in Indiana and possibly Kentucky appears of little consequence in regard to the production of aflatoxin. It is possible, however, that a combination of circumstances: damage, abundant inoculum, and an efficient vector could result in a serious outbreak, particularly in southern Indiana."

Although most corn in the United States is harvested free or nearly free from aflatoxin contamination, the widespread distribution of A. flavus in the corn belt might pose the threat of at least occasional years when conditions are suitable for aflatoxin development in the field. Preliminary information on the 1975 corn crop indicated that the incidence of aflatoxin was much higher than normal. We do not know exactly what factors determine the extent of A. flavus growth and development in the field. Its geographic distribution in corn and the fact that it affects such crops as cotton and peanuts are strong indications that warm climates and/or high temperatures are involved.

In cotton, various factors have been associated with the amount of seed contamination by A. flavus and aflatoxin. Marsh and Taylor [19] thought A. flavus was abundant only in the areas with extremely high temperatures during the time the bolls are opening. Ashworth and co-workers found that cotton bolls in southern California were commonly infected but those in the San Joaquin Valley were not [1]. They characterized the environment of the valley area as having high summer temperatures and low relative humidity. The low humidity was considered the critical factor, causing the bolls to open and dry quickly; higher humidity in Southern California was thought to keep the bolls in a vulnerable condition for a longer time. Other research has shown that low temperatures during boll opening may limit aflatoxin development [10]. Insects may provide openings and/or inoculum for A. flavus infection [3, 30], and the manner in which bolls open may also influence infection [2].

Apparently many factors, physical and biological, affect the epidemiology of A. flavus in cotton. The specific limiting factor probably varies from season to season and from place to place. The same variability probably applies to peanuts, corn, and other crops.

Inoculum levels, pod openings, microbial competition, temperature, and moisture are among the factors which affect the amount of infection in peanuts [4, 11]. Although A. flavus is generally present in the soil and to a limited extent on maturing pods, most fungal growth and aflatoxin accumulation occur after the peanuts are dug and before they are dried. Contamination is likely to be heaviest in peanuts with damaged pods and those which are not dried quickly.

Aspergillus flavus is not known for its ability to invade intact, active plant tissues. Many fungi grow only on living plants, but A. flavus is more appropriately classified as a saprophyte, or a decomposer. Its pathogenic activities in plants and animals cannot be overlooked, but they are probably incidental to the overall survival and well-being of the species. Evidence from

various crops including corn [9, 31] shows that A. flavus occurs primarily in
damaged or relatively inactive tissues, where it must compete with many other
nonspecific saprophytes. A. flavus becomes established or dominates when
conditions are relatively more favorable for it than for other competing fungi
or bacteria. The ability to grow at high temperatures is probably critical in
many situations involving A. flavus. We take advantage of this ability in
selectively isolating A. flavus from grain containing high populations of other
fungi. When dilution cultures of such material are incubated at 40°C, very few
other organisms appear on the culture plates. The same principle might
operate in an ear of corn, a cotton boll, or a grain bin.

5.4   Contamination by Mycotoxins
       during Storage

Although mycotoxins may contaminate food and feedstuffs before harvest, many
of the toxigenic fungi are primarily storage fungi. Species of Aspergillus and
Penicillium in particular are capable of growing rapidly in grain, nuts, and
other products in storage if moisture contents are high enough [6]. Some
storage fungi can grow at relative humidities as low as 65 to 70%; at 85% relative humidity, many species can grow rapidly. The common "field fungi" such
as species of Alternaria, Fusarium, and Cladosporium require very high
relative humidities or moisture contents, and are generally not competitive in
storage situations.
    The growth of Aspergillus flavus along with the production of aflatoxin may
be used as an example of mycotoxin development during storage. Schroeder
[26] has stated that aflatoxin accumulation in most crops, including peanuts,
occurs after harvest. Cottonseed was considered a possible exception. Low
levels of preharvest contamination by A. flavus and other storage fungi are
common, but inoculum may also come from the air, soil, storage and handling
equipment, and so on. The critical factors, then, in determining whether or
not fungi will grow in stored crops are moisture content and temperature.
    Most corn in the United States is field-shelled at moisture contents too high
for safe storage. When harvested during warm weather, corn is extremely
perishable and can quickly be contaminated with aflatoxin. In addition to suitable temperature and moisture, initial fungal population might be an important
factor in determining which fungus or fungi will dominate. Our own observations of freshly harvested corn in Kansas over several years indicate that
A. flavus is more prevalent in some years than in others. This agrees with
other reports cited earlier.
    We tested corn for the presence of various fungi prior to storage tests in
1970 and 1971 [7]. A. flavus was common both years on nonsurface-sterilized
kernels. In surface-sterilized kernels, A. flavus was present in 40 of 2400 in
1970, but in none of 800 kernels in 1971. Those two lots of corn were very
similar in initial moisture content, temperature, and extent of mechanical
damage, but they behaved quite differently in storage. The 1970 corn heated
faster and developed higher populations of storage fungi (primarily A. flavus
and A. niger) than the 1971 corn. During the relatively short time the corn was

at temperatures suitable for A. flavus growth, a maximum of 72 h, up to 80 ppb of aflatoxin was produced in the grain in 1970. Not more than 10 ppb was detected in any samples of the 1971 corn. The data suggest that initial inoculum levels might have accounted for the difference in the two tests.

Aspergillus flavus requires warm conditions and relative humidities above 85% for growth and toxin formation. In corn, this corresponds to moisture contents above 16%. The moisture contents below which fungi cannot grow are different for different commodities, but generally correspond to relative humidities of 65 to 70%. In starchy cereal seeds such as corn, wheat, and sorghum, moisture contents of 13 to 14% are the maximum for safe storage; in seeds with higher oil content, the corresponding moisture contents are lower. Soybeans should be stored below 12.5% and flaxseed below 9%. Although mycotoxins might not be produced at the minimum moisture level which allows fungal growth, storage at such a borderline level should be considered risky. Spoilage losses and mycotoxin production can occur in such grain for many reasons.

A common cause of unexpected spoilage in stored grain is that the actual moisture content of the grain is not known. The moisture testing method or instrument may have been faulty; the operator could have made a mistake; the sample tested may not have been representative; moisture may not be uniform throughout the bulk; and moisture content can increase during storage. Any of those factors can result in estimates of moisture contents that are lower than the true levels. Moisture increases may result from leaks which permit rain to enter or snow to blow in, or from the respiration of insects or fungi in the grain. Temperature gradients are common in bins of grain, and produce moisture migration toward the cooler grain. Such localized moisture increases favor mold growth and mycotoxin development.

If the moisture content of a bin of grain has been determined accurately, the approximate safe storage time for that grain at different temperatures can be predicted. A U.S. Department of Agriculture bulletin [35] lists some estimated safe storage times for corn (Table 1). The figures are based on rates of carbon dioxide evolution from corn and are the length of time which should limit loss of dry matter to 0.5%. Such data show the importance of moisture and temperature in determining rate of spoilage, but do not necessarily indicate mold- or mycotoxin-free storage times. Different lots of grain may behave differently because of differences in mechanical damage, equilibrium moisture content, initial fungal populations, previous storage, and the like.

The following suggestions should help ensure that grain in storage is kept free from excessive mold and mycotoxins: Be sure all the grain is at moisture contents too low for the growth of storage fungi. Be sure of the accuracy of moisture measurements. Inspect the grain regularly, preferably once a week. Weekly inspection should include temperature measurements and a thorough search for insects, mustiness, and wet spots. Any temperature increase that cannot be explained by weather conditions should be considered an indicator that a problem is developing. Regular inspection and common sense can virtually eliminate serious spoilage in grain. Remedial action is most effective and losses are minimal when problems are detected early.

Table 1   Maximum Time for Storage of Shelled
Corn at Various Corn Moistures and
Air Temperatures[a]

Storage air temperature (°F)	Storage time (days) of corn moisture content of:			
	15%	20%	25%	30%
75	116	12.1	4.3	2.6
70	155	16.1	5.8	3.5
65	207	21.5	7.8	4.6
60	259	27.0	9.6	5.8
55	337	35.0	12.5	7.5
50	466	48.0	17.0	10.0
45	725	75.0	27.0	16.0
40	906	94.0	34.0	20.0
35	1140	118.0	42.0	25.0

[a]The times given are those above which mold growth
will cause enough loss in corn quality to bring about
a lowering of grade.
Source: U.S. Department of Agriculture [35].

Where feasible, storage bins should be equipped with fans for aeration.
Aeration can be used to lower grain temperatures, and to eliminate temperature gradients in the bin. Maintaining uniform temperatures prevents
moisture migration. Design specifications and operating guidelines for bin
aeration are available from agricultural experiment stations, extension agents,
and dealers who sell grain bins and related equipment. Many farmers and
others even use aeration fans as an inspection aid. During winter months, for
example, when the fan has not been running, turning it on aids in checking the
odor and temperature of the grain.

5.5   Grain Drying

In recent years most grain harvested too wet for safe storage has been dried
in various kinds of heated air dryers, usually at high temperatures. Such
drying requires large amounts of fuels such as natural and LP gas, which are
becoming more expensive and less available. Crop drying may be considered
a low priority use for our remaining stocks of petroleum products. The fuel
shortage has increased interest in alternatives to conventional grain drying.
    A common initial reaction to the grain drying problem is to suggest that
the crop simply be left in the field to dry naturally, and it is true that sooner
or later the grain will dry on its own. However, it is also true that the longer

a crop is left in the field after it reaches its maximum dry matter content, the lower the yield will be. Field losses attributed to insects, fungi, birds, lodging, shattering, etc., are a high price to pay for natural field drying. The bushels of corn saved by early harvesting may more than offset the cost of artificial drying. Also, leaving grain in the field increases the chances of harvest being delayed by wet weather. Growers may have to wait until winter or spring to complete the harvest. As was pointed out earlier, such conditions contribute to field production of mycotoxins.

In view of the advantages of early, high moisture grain harvesting, and the increasing limitations on using fuel to dry grain, alternative grain handling systems must be used. The alternatives include increasing the efficiency of high temperature dryers, drying at low temperatures, and high moisture storage.

For low temperature drying in a storage bin, unheated or slightly heated air is forced through the grain [29]. In most systems airflow is upward through a perforated metal floor. The drying process requires several days or weeks, depending primarily on the quantity of grain and the volume of air that can be moved through it. The final moisture content is determined by the temperature and humidity of the incoming air. One of the chief hazards of low temperature drying is that all the grain does not dry simultaneously; rather, a drying "front" gradually moves through the grain. Grain below the front or drying zone is dry; the grain above is at approximately the original moisture content. The grain in the top of the bin may or may not dry soon enough to avoid spoilage. Depending on initial grain moisture, amount of mechanical damage, uniformity or airflow, and average temperatures, some grain could be heavily invaded by storage fungi and contaminated with mycotoxins. The drying front can be broken up or eliminated by periodic mixing or turning of the grain during the drying period. The grain then would have more uniform moisture content throughout, and areas of distinctly wet or dry grain would be eliminated.

Because ambient air during late fall and early winter may have relatively little drying potential, supplemental heat can be added to aid in moisture removal. The most common source has been supplemental electric heat, enough to raise incoming air temperatures just a few degrees. Also, when air is drawn over the fan motor, the motor heat can add one or two degrees to the air temperature. Solar heat can be collected in various ways and used to assist in grain drying. Extensive research currently underway, as well as limited commercial usage, should soon show whether or not "solar grain drying" is practical.

Adding supplemental heat does not shorten the time required to move a drying front completely through the grain. The only ways to shorten the total drying time are to increase total airflow and decrease the amount of grain in the bin. Adding extra heat only results in a lower final moisture content. Low temperature, in-bin drying systems depend on adequate ventilation through all the grain. Fine material and trash, which are extremely susceptible to attack by fungi, tend to accumulate under the filling spout, restricting airflow in this area. Airflow tends to go around accumulations of this material, which does not dry properly, and may become heavily mold invaded. Passing the grain over a simple screen-type cleaning device on the way into the bin greatly reduces that problem.

Slow drying with ambient or slightly heated air seems to be energy efficient, but there are risks, as indicated. That type of drying depends on weather conditions, which vary from year to year. A given system may dry corn safely in 8 or 9 years out of 10 on the average, but allow considerable spoilage the other year or two. Low temperature drying requires careful attention and alertness on the part of the operator. When grain is not drying fast enough and spoilage seems likely, an alternative plan should be available, such as immediate use as animal feed, high temperature drying, or treatment with chemical preservatives.

5.6   High Moisture Grain Storage

As an alternative to drying, grain can be stored at high moisture contents (20-30% or more) and used as livestock feed. High moisture grain can be stored under airtight conditions [13] or treated with chemicals such as propionic acid [15, 24]. Properly done, either system protects the grain from mold and mycotoxin problems. Many livestock feeders find that high moisture grain is superior to dry grain in terms of feed efficiency and/or rate of gain. High moisture storage allows feeders to take advantage of high yields from early harvest, and eliminates the need for drying.

Several kinds of structures are used for airtight storage or ensiling. Because many fungi can grow at very low oxygen concentrations, relatively small air leaks in high moisture grain silos can result in extensive spoilage. The most satisfactory silos, in terms of minimum spoilage risk, are the specially designed "glass lined" bolted steel units. They represent a large capital investment, but generally produce a high quality livestock feed with minimal spoilage or fermentation losses.

Upright concrete silos, bunkers, pits, or simple plastic-covered piles are also used for ensiling high moisture grain, but the risk of spoilage and mycotoxin contamination is higher in these kinds of storages. In many cases the "seal" on the silo is merely a layer of material so heavily caked with fungi that oxygen cannot easily get through it. Very rapid fungal growth depletes the oxygen in the main grain bulk, but continues long enough at the surface to form a "seal." A further problem with nonairtight silos is that grain must be removed at a fairly rapid rate in order to prevent the exposed surface from becoming moldy. Warm weather makes this spoilage problem particularly troublesome.

The use of chemical grain preservatives is a recent development in storing high moisture grain. Propionic acid is the most common material used, applied as 100% propionic acid or in mixtures with acetic acid, isobutyric acid, or formaldehyde. Properly treated, grain of almost any moisture content can be stored for at least a year without fungal spoilage. As with ensiled grain, acid-treated grain is suitable only for livestock feed. However, it is not affected by exposure to air, and does not require any particular kind of storage structure.

Table 2   Propionic Acid Required for
Preventing Mold Growth in
High-Moisture Grain

Moisture content (%)	Propionic acid required		
	%	lb/ton	oz/bu
18	0.3-0.6	6-12	2.5-5
22	0.5-0.8	10-16	4.0-7
26	0.6-1.0	12-20	5.0-8
30	0.8-1.2	16-24	7.0-10

Source: Sauer [23].

The amount of preservative required depends primarily on moisture content of the grain, and to a lesser extent on temperature and duration of storage. Table 2 shows the amounts of propionic acid required to preserve grains of various moisture contents [23]. A range of treatment rates is given for each moisture content. The lower rates are probably suitable for short-term storage in cool weather, such as from fall harvest through the winter. The higher rates are recommended for storage up to 1 year. Acid-treated grain should be inspected regularly during storage.

As with any storage system, acid treatment has advantages, disadvantages, and risks. Spoilage can occur if the application rate is too low or not uniform, so that pockets of untreated grain exist. Moisture migration can result in a wet spot where fungi will grow. As with dry grain, aeration can be used to break up temperature gradients that cause moisture migration. Fungi will also grow in acid-treated grain that is in contact with unprotected concrete or steel. Such surfaces should be covered with plastic or coated with acid-resistant paint.

5.7   Summary

No attempt has been made in this section to deal with all crops or all toxigenic fungi. The examples used may be considered as representative of the kinds of problems that exist and how they can be avoided. Other crops and/or fungi could have been used to illustrate various points. Rice, for example, is usually harvested under conditions suitable for aflatoxin development; in colder climates, grains are vulnerable to potentially toxigenic species of Fusarium and Penicillium, many of which can grow at temperatures near the freezing point. Although any specific mycotoxin may be considered as primarily a warm climate problem or a cool climate problem, mycotoxins in general should be viewed as a potential threat to food and feedstuffs throughout the world.

Fortunately, because specific fungi must be present on any given material, and specific environmental conditions must be met, most foods and feeds are only occasionally contaminated with mycotoxins. Some mycotoxins are produced by fungi growing in plants in the field, such as pasture grasses or various grains, oilseeds, and nut crops. Resistant plant varieties, altered cultural practices, and knowledge of conditions under which problems develop are the main defenses that can be used to prevent or avoid field-contaminated crops.

Most of the mycotoxin-producing fungi are adapted to growing on harvested crops during storage. Although many species of fungi and many commodities are involved, mycotoxin production can be prevented simply by making conditions unsuitable for fungal growth. The oldest and most widely used control measure is storing the commodity at a low moisture content. It is imperative that crops harvested with high moisture contents be dried promptly or otherwise conditioned for storage. Equally important, products held in long-term storage must be inspected regularly to detect signs of spoilage before mycotoxin development or other serious losses occur.

References

1. L. J. Ashworth, Jr., J. L. McMeans, and C. M. Brown (1969): Infection of cotton by Aspergillus flavus: Epidemiology of the disease. J. Stored Prod. Res. 5:193-202.
2. L. J. Ashworth, Jr., J. L. McMeans, and C. M. Brown (1969): Infection of cotton by Aspergillus flavus: Time of infection and the influence of fiber moisture. Phytopathology 59:383-385.
3. L. J. Ashworth, Jr., R. E. Rice, J. L. McMeans, and C. M. Brown (1971): The relationship of insects to infection of cotton bolls by Aspergillus flavus. Phytopathology 61(5):488-493.
4. L. J. Ashworth, Jr., H. W. Schroeder, and B. C. Langley (1965): Aflatoxins: Environmental factors governing occurrence in Spanish peanuts. Science 148:1228-1229.
5. P. J. Brook (1969): Pithomyces chartarum in pasture, and measures for prevention of facial eczema. J. Stored Prod. Res. 5:203-209.
6. C. M. Christensen and H. H. Kaufmann (1969): Grain Storage: The role of Fungi in Quality Loss. Univ. of Minn. Press, Minneapolis, 153 pp.
7. H. H. Converse, D. B. Sauer, and T. O. Hodges (1973): Aeration of high moisture corn. Trans. Am. Soc. Agric. Eng. 16:696-699.
8. U. L. Diener, N. D. Davis, and G. Morgan-Jones (1974): Toxigenic fungi from bermudagrass hay. Proc. Am. Phytopathol. Soc. 1:104.
9. B. Doupnik, Jr. (1972): Maize seed predisposed to fungal invasion and aflatoxin contamination by Helminthosporium maydis ear rot. Phytopathology 62:1367-1368.

10. D. E. Gardner, J. L. McMeans, C. M. Brown, R. M. Bilbrey, and L. L. Parker (1974): Geographical localization and lint fluorescence in relation to aflatoxin production in Aspergillus flavus infected cottonseed. Phytopathology 64:452-455.

11. G. J. Griffin and K. H. Garren (1974): Population levels of Aspergillus flavus and the A. niger group in Virginia peanut field soils. Phytopathology 64:322-325.

12. D. Gröger (1972): Ergot, in S. Kadis, A. Ciegler, and S. J. Ajl (eds.): Microbial Toxins. Academic Press, New York, pp. 321-373.

13. M. B. Hyde (1974): Airtight storage, in C. M. Christensen (ed.): Storage of Cereal Grains and Their Products. Am. Assoc. Cereal Chemists, St. Paul, Minn., pp. 383-419.

14. A. Z. Joffe (1971): Alimentary toxic aleukia, in S. Kadis, A. Ciegler, and S. Ajl (eds.): Microbial Toxins, Vol. 7. Academic Press, New York, pp. 139-189.

15. G. M. Jones, D. N. Mowat, J. I. Elliot, and E. T. Moran, Jr. (1974): Organic acid preservation of high moisture corn and other grains and the nutritional value: A review. Can. J. Anim. Sci. 54:499-517.

16. S. Kadis, A. Ciegler, and S. Ajl, eds. (1971): Microbial Toxins, Vol. 7, Academic Press, New York, 401 pp.

17. B. Koehler (1959): Corn ear rots in Illinois. Univ. of Ill. Agric. Exp. Sta. Bull. 639. 87 pp.

18. E. B. Lillehoj, W. F. Kwolek, D. I. Fennell, and M. S. Milburn (1975): Aflatoxin incidence and association with bright greenish-yellow fluorescence and insect damage in a limited survey of freshly harvested high-moisture corn. Cereal Chem. 52:403-412.

19. P. B. Marsh and E. E. Taylor (1958): The geographic distribution of fiber containing fluorescent spots associated with Aspergillus flavus in the United States cotton crop of 1957. Plant Dis. Rep. 42:1368-1371.

20. L. E. Melchers (1956): Fungi associated with Kansas hybrid seed corn. Plant Dis. Rep. 40:500-506.

21. S. B. Puranik and D. E. Mathre (1971): Biology and control of ergot on male sterile wheat and barley. Phytopathology 61:1075-1080.

22. G. W. Rambo, J. Tuite, and R. W. Caldwell (1974): Aspergillus flavus and aflatoxin in preharvest corn from Indiana in 1971 and 1972. Cereal Chem. 51:595-604.

23. D. B. Sauer (1973): Grain preservatives for high-moisture feed grains. Report from the U.S. Grain Marketing Research Center, Manhattan, Kansas, 9 pp.

24. D. B. Sauer and R. Burroughs (1974): Efficacy of various chemicals as grain mold inhibitors. Trans. Am. Soc. Agric. Eng. 17:557-559.

25. V. F. Savel'yev (1962): The characteristics of the microflora of corn cobs in the southern part of the Ukraine. Mikrobiol. Zhur. Akad. Nauk. Ukrain. R.S.R. 24(2):39-44.

26. H. W. Schroeder (1969): Factors influencing the development of aflatoxins in some field crops. J. Stored Prod. Res. 5:187-192.

27. H. W. Schroeder, R. J. Cole, R. D. Grigsby, and H. Hein, Jr. (1974):
     Inhibition of aflatoxin production and tentative identification of an aflatoxin
     intermediate "versiconal acetate" from treatment with dichlorovos.
     Appl. Microbiol. 27:394-399.

28. O. L. Shotwell, C. W. Hesseltine, and M. L. Goulden (1973): Incidence of
     aflatoxin in southern corn, 1969-1970. Cereal Sci. Today 18:192-195.

29. G. C. Shove (1973): New techniques in grain conditioning, in R. N. Sinha
     and W. E. Muir (eds.): Grain Storage: Part of a System. The AVI Publ.
     Co., Westport, Conn., pp. 209-228.

30. L. W. Stephenson and T. E. Russell (1974): The association of Asper-
     gillus flavus with hemipterous and other insects infesting cotton bracts
     and foliage. Phytopathology 64:1502-1506.

31. J. J. Taubenhaus (1920): A study of the black and yellow molds of ear
     corn. Texas Agric. Exp. Sta. Bull. No. 270. 38 pp.

32. J. Tuite (1959): Low incidence of storage molds in freshly harvested seed
     of soft red winter wheat. Plant Dis. Rep. 43:470.

33. J. Tuite (1961): Fungi isolated from unstored corn seed in Indiana in
     1956-1958. Plant Dis. Rep. 45:212-215.

34. J. Tuite and R. W. Caldwell (1971): Infection of corn seed with Helmintho-
     sporium maydis and other fungi in 1970. Plant Dis. Rep. 55(5):387-389.

35. U.S. Department of Agriculture (1968): Guidelines for mold control in
     high-moisture corn. Farmers' Bull. No. 2283, 16 pp.

36. S. G. Yates (1971): Toxin-producing fungi from fescue pasture, in S.
     Kadis, A. Ciegler, and S. Ajl (eds.): Microbial Toxins, Vol. 7.
     Academic Press, New York, pp. 191-206.

David B. Sauer

PART 6

REGULATORY ASPECTS OF THE MYCOTOXIN PROBLEM IN THE
UNITED STATES

6.1 Introduction

The Federal Food, Drug, and Cosmetic Act, which embodies the nation's concern for the safety and quality of foods, drugs, and cosmetics, contains a section [402(a)1] which defines a food as adulterated "if it bears or contains any poisonous or deleterious substance which may render it injurious to health" [1]. The Federal Food and Drug Administration (FDA) has the authority to enforce the Act and can remove from interstate commerce any food or feed found to be adulterated.

Traditionally mold contamination of food has been considered a violation of another section of the Act [402(a)3] which defines a food as adulterated "if it consists in whole or in part of any filthy, putrid, or decomposed substance" [2]. However, toxic mold metabolites (the mycotoxins which are the subject of this book) when present in foods, even in the absence of obvious mold growth, are treated under Section 402(a)1. The courts have upheld this interpretation for aflatoxin-contaminated foods [3].

Although the law requires that the FDA remove from interstate commerce foods found to be adulterated, effective regulatory control entails far more than implementation of the legal processes detailed in the Act. The shape of a total control program depends on the type of contamination problem to be attacked, since each, including the mycotoxin problem, has its own unique characteristics. These total programs are very often dynamic in nature since the scientific and technological knowledge which provides the basis for any sound control scheme is itself continuously changing.

This section will describe the current (1974) features of various mycotoxin control programs and will outline how these programs evolved to their present status. It will also give some account of how current investigations, both within FDA and within the rest of the scientific community, may be used to frame future programs. Before entering into this discussion, however, it will be necessary to examine some general features of the mycotoxin contamination problem which must be considered in designing and implementing proper control programs.

6.2 General Characteristics of the Problem of Mycotoxins in Foods

In some respects the mycotoxin problem is similar to, and in other respects it is distinct from, the problems associated with other chemical contaminants of food. The following list of generalizations contains some of the important features of the mycotoxin problem, without providing elaboration on these similarities and distinctions. Although some of these generalizations may seem patently obvious, a systematic control approach depends on an appreciation of

each of these characteristics of the problem. Much of the information in the
list derives from over 10 years' experience with aflatoxin problems, but most
is surely applicable to other mycotoxins.

1. Mycotoxin contamination of foods can occur on growing plants in the
   field, but other major sources are faulty or inadequate harvesting and
   and storage practices.
2. The observation of growth of a known mycotoxin-producing mold on a
   food is not evidence that its toxic product(s) is present in the food.
   Conversely, the failure to observe visible growth of a mycotoxin-
   producing mold on a food cannot be taken as proof that its toxic
   product(s) is absent from the food. Therefore, methods for detecting
   and measuring mycotoxins in foods must be directed at the mycotoxins
   themselves, and not at the producing molds.
3. With the exception of liquid products, mycotoxin contamination of foods
   is often highly localized within a lot. Sampling of a lot is usually the
   major source of error in the assay result obtained for a lot.
4. Mycotoxins are relatively stable organic compounds of diverse struc-
   tural types. Most can survive the common food-processing operations.
5. In the United States, levels of mycotoxins which might be associated
   with acute (immediately observable) toxic effects in humans are not
   encountered in human food, but may be observed in animal feed.
   Normal agricultural grading practices remove heavily molded products,
   which might contain such levels of mycotoxins, from human food
   channels. Of greater concern to humans are the possible effects of
   relatively long-term and low level ingestion (subacute and chronic
   toxicity effects such as mutagenesis, teratogenesis, and carcinogenesis). Almost always, these effects can be gauged only from
   studies in experimental animals. Such studies are inevitably lengthy
   and very costly, and although their meaning relative to human health
   is not always clear, they must be relied upon for establishing safety
   guidelines.
6. Mycotoxins have been found in animal feed at levels which can produce
   acute toxic effects in farm animals in the United States; but always of
   additional concern is the question of residues of the mycotoxins or
   their toxic metabolites in meat, milk, and eggs.

Finally, there are two points which often cause confusion and which should
be clarified. First, even though mycotoxins are natural in origin, they are, if
present in food, added substances (i.e., they are not natural components of
foods). Second, although mycotoxins are added substances, it is recognized
that their presence in food is not always avoidable (i.e., current agricultural
practice cannot ensure the complete absence of mold growth on food, and
current processing and manufacturing practices cannot accomplish the
complete removal of contaminated portions from a lot of food).

6.3   Evolution of the Current Aflatoxin
          Control Program

6.3.1   Background

The possibility that mold-contaminated peanuts could contain a highly toxic compound came to the attention of the FDA in the early 1960s, soon after the outbreak of the now famous "turkey X" disease in England (see Sec. 3.6.1). Most of the early information derived from personal exchanges between government scientists from that country and the United States. The most significant impetus to control-directed activity on this problem came with the publication demonstrating that these toxicants, now known to have been the aflatoxins (Sec. 2.1), were potent hepatocarcinogens in experimental animals [4], and with the surveys which showed a significant incidence of aflatoxins in peanut products in the United States [5].

Clearly the presence of a carcinogen in any food represented an extremely undesirable situation, but it was just as clear that the means for control of the problem were not readily available or even well understood. In the absence of specific information regarding possible safe tolerances for aflatoxins or relating to the question of the extent to which aflatoxin contamination might be limited by the best available and practical agricultural and processing technology, in 1965 the FDA announced an action guideline of 30 ppb total aflatoxins. In 1969 this action guideline was reduced to its current level of 20 ppb. This guideline is subject to change as new information, relating either to the toxic effects of the aflatoxins or to the degree to which their presence in foods can be avoided by the best available and practical agricultural and processing technology, is developed and becomes sufficient in scope and content to allow reconsideration of the current guideline.

6.3.2   Analytical Methodology

It was recognized quite early that the development of sampling plans and assay methods to detect, measure, and confirm the presence of aflatoxins in foods (see also Secs. 2.1.7 and 3.5.2.5) was crucial to any attempt to control the problem [6]. Several scientific societies have taken up the difficult chore of coordinating and discriminating among the various analytical methods which have appeared in the literature. The major reason for these efforts is to assure the reliability of methods by subjecting them to interlaboratory collaborative studies. Valid sampling procedures and analytical methods are, obviously, of supreme importance and any evaluation of experimental, surveillance, or regulatory analyses must take into account the reliability of the sampling procedures and the methods used in collecting the raw data.

At present, methods for aflatoxins and other mycotoxins are under study and review by the American Oil Chemists' Society (AOCS), the American Association of Cereal Chemists (AACC), the Association of Official Analytical

Chemists (AOAC), and the International Union for Pure and Applied Chemistry (IUPAC). The AOAC has adopted a greater number of methods than the other societies, and does this through a system of specialists called associate referees who conduct interlaboratory studies, prepare annual reports of their activities, and submit these reports to the society's coordinator, or general referee for mycotoxins. The general referee prepares an annual report which is submitted to the Association for review [7]; in this report the general referee makes recommendations on each of the specific areas, and these are either accepted or rejected by the Association. Mycotoxin methods adopted by the AOAC are published in a single chapter of the society's publications, the Official Methods of Analysis [8]. These are usually the methods used by the FDA in its regulatory programs. The other concerned societies also publish adopted methods. The goals of all four societies are coordinated by formal intersociety committees.

The FDA and all other governmental and industrial groups concerned with mycotoxin control necessarily have deep involvement with these societies and their activities. The development of valid analytical methods and sampling plans is central to the implementation of any type of control system.

6.3.3   Control of Specific Aflatoxin-
         Susceptible Commodities

After the discovery of the problem of aflatoxin contamination of peanuts in the United States, the FDA and the U.S. Department of Agriculture (USDA) began research and surveillance programs to learn if other commodities were subject to contamination. Indeed, it was not long before cottonseed, corn, Brazil nuts, copra, various domestic tree nuts, and pistachio nuts were added to the list of susceptible commodities. Through a series of continuing surveillance programs, the two government agencies have gathered information on a wide range of commodities and have been able to demonstrate that, in contrast, a number of other important foods produced in this country are at least not highly susceptible to aflatoxin contamination.

Thus, the major control efforts and regulatory activities are now directed at the susceptible foods mentioned. Further surveillance of other foods for aflatoxin contamination susceptibility is continuing. If a commodity is found to be susceptible and the degree of susceptibility is high, then the FDA will usually seek to have implemented, on a voluntary basis, control programs at the farm, shipping, or processing level. Most often the USDA is involved in these programs. The FDA then maintains a regulatory program aimed at finished products. Some of the programs discussed in the following sections are of this type. If the evidence is that the contamination of a particular commodity is occasional and not at all usual, then the FDA relies simply on its regulatory program to uncover such an occurrence and will act to remove the contaminated item from interstate commerce.

6.3.3.1   Peanuts

After the recognition of the susceptibility of peanuts to aflatoxin contamination, the USDA and the industry established a marketing agreement which included a plan for sampling, analysis, and certification of all shelled peanuts destined for human consumption. This agreement, which is still fully in effect, is designed to remove lots of aflatoxin-contaminated raw peanuts prior to finished product manufacture. The FDA is involved in this program on an advisory basis, and receives yearly, for evaluation, the USDA reports of analyses of raw shelled peanuts. Laboratories which carry out the aflatoxin analyses must be certified as competent by the Peanut Administrative Committee.

Within the context of the rather complicated sampling plan, which is based on certain assumptions about the distribution of contamination, lots of raw peanuts received a "negative" certification if the analytical value found for the lot is less than 25 ppb [9]. The FDA does not object to the interstate shipment of peanuts carrying such a certificate. Finished product manufacture further reduces the aflatoxin content of peanuts.

Direct FDA sampling and analysis are aimed at finished peanut products, and, at present, legal action is taken against finished products containing greater than 20 ppb total aflatoxins.

6.3.3.2   Brazil Nuts and Pistachio
                 Nuts

The discoveries that pistachio nuts and in-shell Brazil nuts were highly susceptible to aflatoxin contamination prompted FDA action on these important commodities. Voluntary importer control programs have been set up; these programs call for sampling and analysis of every lot of these products before they are allowed entry into the United States. As is the case with peanuts, the role of the FDA has been advisory. The plans for sampling and analysis have been reviewed and approved by the FDA. The USDA carries out the testing and submits results to the FDA for evaluation. At any time the FDA can ask for changes in the nature of the testing program if such changes are warranted.

FDA and USDA scientists have been involved in a number of consulting visits to the exporting countries to aid those nations in setting up their own control programs.

6.3.3.3   Corn

The USDA and the FDA have carried out a number of surveys to determine the incidence of aflatoxins in corn. At present, these surveys indicate that contamination is most likely to occur in the southeastern states, where climatic and agricultural conditions are most conducive to Aspergillus flavus (Sec. 1.2.3.3) growth. However, some contamination has been observed in the south central and midwestern states.

Nearly 90% of the corn is sold directly from the farm for use as animal feed. Some of the remainder is used for making cornmeal, breakfast cereal, and grits. Manufacture of wet process products and alcoholic beverages is another major use. Fresh, sweet corn, like other fresh vegetables, is not susceptible to this type of contamination.

The FDA is now working with members of the corn industry and the USDA to coordinate and collect technical information about the problem. Of particular importance is the establishment of reasonable plans for sampling and analysis of corn; the ultimate goal is some form of testing program for corn prior to marketing. Simply because of the volume of the product, the development of such a program is beset with some rather severe problems. As with other commodities, adequate sampling schemes are most difficult to devise.

Until such a program is fully functioning, the FDA will continue to monitor corn products to ensure the protection of the consumer.

6.3.3.4   Cottonseed

Aflatoxin contamination of cottonseed appears to be concentrated in the southwestern part of the United States, although other areas can be affected. The problem seems to arise because of bollworm infestation combined with the practice of crop irrigation.

Cottonseed and the meal derived from it have been subject to some rather high levels of aflatoxin contamination. These products are used for animal feed and not for direct human consumption. At present the FDA monitors interstate shipment of these products. Some affected states, particularly California, have instituted successful control programs for the large portion of the crop which is involved only in intrastate commerce.

6.3.3.5   Tree Nuts

Liaison programs have been established with pecan, almond, and walnut industry associations. The problem with these commodities does not appear to be a severe one, but attempts are being made to set up voluntary control programs.

6.3.4   Aflatoxin Residues in Food-
Producing Animals

The major concern the FDA has with aflatoxin-contaminated animal feed revolves around the question of aflatoxin residues in edible tissue, milk, and eggs. The FDA is continually seeking to learn the maximum level of aflatoxin in feed which will result in no detectable aflatoxin in the edible animal products. Until such information is obtained (and is obtained in a highly systematic way, with the resulting data susceptible to unambiguous interpretation) the current guideline of 20 ppb will be maintained for animal feed ingredients.

6.4   Research Activities Related to
       Mycotoxin Control

Although a great deal has been learned about aflatoxin contamination of foods,
there is still much to be discovered. Following is a brief list of activities now
underway, to varying degrees, in the USDA, the FDA, and affected industries.
The emphasis of all FDA research on aflatoxins, and on other mycotoxins as
well, can be summarized as problem identification and establishment of
enforcement guidelines. The research work of the USDA and affected industries
is usually in the general area of prevention of contamination and "cleanup" of
contaminated lots of food.

1. Determination of the agricultural practices which must be controlled
    to prevent contamination.
2. Search for mold-resistant plant varieties.
3. Development of plans for sampling commodities for contamination.
4. Development of new analytical methods and improvement of existing
    methods.
5. Development of reliable rapid screening methods.
6. Establishment of techniques for removing contaminated portions from
    a lot of food.
7. Development of procedures for detoxification of contaminated products.
    (Some interesting work has been carried out by the cottonseed industry
    and the USDA in this area. Ammoniation of cottonseed meal for
    purposes of aflatoxin detoxification has yielded some promising
    results. The testing of this product for its nutritional characteristics
    and its safety is now being conducted by the industry and the USDA.
    Similar treatment of corn also appears to carry some promise.)
8. Determination of the relationship between levels of contaminant in
    animal feed and levels in edible tissue, milk, or eggs.
9. Determination, through studies with experimental animals or from
    epidemiological analyses, of the toxicological importance of these
    contaminants for humans.
10. Determination of the toxicological effects of mycotoxins in farm
    animals.
11. Establishment of the chemical and physical characteristics of toxic
    fungal metabolites which are potential food contaminants and which are
    as yet still of unknown chemical structure.

Although these 11 research activities on mycotoxins are not the only ones
of scientific interest, they bear most directly on the question of control of the
problem. As a matter of purely scientific interest, the mycotoxin problem
has many attractive features and in addition can serve to bring university
scientists, for instance, quite closely into the realm of public service. A
total attack on the problem will bring together agronomists, agricultural
engineers, plant pathologists, plant geneticists, mycologists, organic chemists,
analytical chemists, biochemists, toxicologists, veterinarians, food technolo-
gists, and statisticians.

6.5   Development of Programs for
       Other Mycotoxins

The FDA, after its initial experience with aflatoxin, began to devote resources
to an examination of some of the other mycotoxins. The FDA activities on
these other mycotoxins have until now been primarily of an investigatory
nature (i.e., research or analytical methods, confirmatory tests, isolation
and purification of mycotoxins, review of the toxicological literature, and
field surveillance programs).

In choosing mycotoxins for investigation, the usual approach is to review
the toxicological literature to determine just how much can be estimated about
the health hazard potential for animals and humans which might be expected if
the mycotoxin were to be found in the food or feed supply. In almost every
instance, such a literature search allows some rough estimation of the possible
effects in farm animals (much of the early work on mycotoxins is to be found in
the veterinary literature). The classification of a mold metabolite as a mycotoxin results from some type of toxicological study, usually an acute toxicity
study in a mammalian species. What is usually absent from the literature is
work on the effects of mycotoxins when administered orally to experimental
animals in a subacute or chronic fashion. Since concern with the acute effects
of mycotoxins in humans in the United States is of minimal interest, it is
necessary to obtain the missing toxicological information. However, studies
to collect this information are extremely costly. Therefore, such studies are
conducted only if there is evidence that the mycotoxin can occur as a food or
feed contaminant under natural (i.e., field as opposed to laboratory) conditions.

To collect data on contamination, valid analytical methods and analytical
standards must be available. The FDA maintains 18 district laboratories
throughout the country to carry out surveillance activities of this type, in
addition to their regulatory activities. Considerable surveillance activity has
also been carried out by the USDA and by public health institutions throughout
the world. The results of all these activities to date have recently been com-
piled by Stoloff [10].

If a mycotoxin is found to have a significant incidence in foods, appropriate
toxicological investigations will begin. The result may be the establishment of
some guideline or tolerance for the mycotoxin in food or feed. Toxicological
information is also used in conjunction with data on food consumption patterns,
and the question of the extent to which a contaminant might be unavoidable
must also enter into the final regulatory decision.

Currently the FDA has under active study the ochratoxins, patulin, sterigmatocystin, penicillic acid, zearalenone, and the trichothecenes (particularly
T-2 toxin). All of these mycotoxins except sterigmatocystin have been detected
in some item of food in the United States [10], but, as has been mentioned,
there is at present insufficient toxicological information from which to derive
an assessment of the significance to human or animal health of the levels which
ave been found. The FDA is currently studying the toxicology of patulin,
ralenone, and T-2 toxin; plans for similar studies on penicillic acid and
toxin have been made. Information is also being collected on the extent of

transmission of these toxins to the edible tissues, or eggs of farm animals receiving these compounds in their diets.

As soon as sufficient information is available, the FDA will move to establish guidelines and take regulatory action where necessary. Plans for developing other aspects of control activity are already underway for these other mycotoxins.

6.6   Future Trends

The outline of the regulatory, or control, aspects of the mycotoxin problem traced here describes the situation as it now exists. Although the FDA must apply its authority in a consistent and fair way, this does not mean that the Agency's policies are unchanging. New scientific and technological information is always becoming available, especially in a relatively new field of study such as mycotoxicology. (This is not equally true of all mycotoxins. For instance, further toxicological studies on the carcinogen aflatoxin are not likely to reveal new information which would cause the FDA to change its position on this mycotoxin; with regard to the question of the extent to which the occurrence of aflatoxin in peanut products is unavoidable, there appear to be no new agricultural or processing developments which are likely to change the situation in the near future.)

The mycotoxin problem, unlike problems with some other food contaminants, appears to be without national boundaries. Investigations reveal that aflatoxin contamination of food is worldwide in occurrence and seems to be heavily concentrated in those areas in which climate and agricultural conditions are most suitable to mold growth: the humid, tropical regions.

Further investigations would no doubt reveal the occurrence of many more toxic mold metabolites in foods. The discovery of mycotoxins in foods in the United States, which has one of the world's most sophisticated agricultural and food marketing and processing systems, probably gives only a slight indication of the extent of the problem on a worldwide basis. The central question becomes: to what extent are mycotoxins involved in idiopathic disease in man? The trail of research leading to an answer to this is long and very difficult.

Even in the absence of a complete answer to that complex question, the simple recognition that mold contamination of food can be more than a simple aesthetic problem should force us, as a nation, to attempt to eliminate those practices which can lead to such contamination. Much is known about how food commodities should be harvested, dried, stored, and transported to avoid mold growth, and those in charge of each of these activities should be made aware of the need to ensure that each of these practices is carried out in a sound and responsible way. Although thorough application of the best available agricultural and manufacturing practices will probably not completely eliminate the problem, it will certainly go a long way to ensure that mycotoxin contamination will not take place.

On a worldwide basis the problem is more complex, since the discovery of adulterated food often means that the food must be destroyed. Clearly, if the alternative is starvation or famine, destruction of food, even contaminated food, becomes a serious moral question. One hopes that such a question will not have to be faced in the United States, but if severe food shortages do occur, the decision on what course of action must be taken, even on contaminated food, is a societal decision which would go far beyond the scope of the regulatory concepts discussed in this section.

6.7   Recent Developments

In the time since this article was written (1974), the FDA has published a proposal to establish a tolerance of 15 ppb for total aflatoxins in consumer peanut products. The basis for the proposal is detailed in the preamble to the proposed regulation and can be found in the Federal Register [11]. Public comment on the proposed regulation was solicited, has been evaluated, and a final order on this regulation is expected to be published by the time this book appears. Further information on recent FDA regulatory policy on mycotoxins can be found in the Proceedings of the recently held Conference on Mycotoxins in Human and Animal Health [12]; the Proceedings should be available by the end of 1977.

Further developments on sampling plans and collaboratively studied methods of analysis for aflatoxins can be found in the article by Schuller et al. [13].

International interest in the economic and public health problems associated with mycotoxin contamination of foods and feeds has grown considerably in the past 2 years. The Food and Agriculture Organization (FAO) of the United Nations, the World Health Organization (WHO), and the United Nations Environment Program (UNEP) are undertaking a review of the entire problem; the results of this review and FAO/WHO/UNEP assistance programs will be the subject of an international conference scheduled to take place in Nairobi in the fall of 1977.

References

1. Federal Food, Drug, and Cosmetic Act, Chapter IV, 402(a)1, U.S. Govt. Printing Office, Washington, D.C.
2. Federal Food, Drug, and Cosmetic Act, Chapter IV, 402(a)3.
3. United States versus an Article of Food — White Corn, etc.: United States District Court for the District of Kansas, Civil Action No. T-4173, Order Filed January 22, 1971.
4. M. C. Lancaster, F. P. Jenkins, J. McL. Philp, K. Sargeant, A. Sheridan, and J. O'Kelly (1963): Nature (London) 192:1095.
5. R. W. Detroy, E. B. Lillehoj, and A. Ciegler (19): In A. Ciegler, S. Kadis, and S. J. Ajl (eds.): Microbial Toxins, Vol. 6. Academic Press, New York, pp.   .
6. L. Stoloff (1972): Clin. Toxicol. 5:465.
7. The most recent of the series is L. Stoloff (1974): J. Assoc. Off. Anal. Chem. 57:274.

8. Official Methods of Analysis (W. Horwitz, ed.), Association of Official Analytical Chemists, P.O. Box 540, Benjamin Franklin Sta., Washington, D.C., 11th ed., 1970.

9. Voluntary Code of Good Practices for Peanut Product Manufacturers, National Peanut Council, 10th ed., 1972.

10. L. Stoloff, Division of Agricultural and Food Chemistry, 168th Meeting, American Chemical Society, Atlantic City, New Jersey, September 1974. This paper is part of a symposium on mycotoxins; the symposium will be published, sometime in 1975, in the Advances in Chemistry Series of the ACS.

11. Federal Register, Vol. 39, December 6, 1974, pp. 4278-42752.

12. Proceedings of the Conference on Mycotoxins in Human and Animal Health, College Park, Maryland, October 4-8, 1976.

13. P. L. Schuller, W. Horwitz, and L. Stoloff (1976): J. Assoc. Off. Anal. Chem. 59:1315.

Joseph V. Rodricks

GLOSSARY

The glossary includes selected mycological terms commonly used by mycologists and plant pathologists and medical terms commonly used by veterinarians and physicians. Definitions are brief and therefore not always all inclusive, but they are listed to aid scientists not familiar with common terminology in other disciplines and to assist lay persons having peripheral interests to the mycotoxicological subject matter discussed in this handbook.

acanthosis: diffuse hyperplasia with thickening of the prickle-cell layer of the epidermis

acervulus: a cushion-like mass of hyphae having conidiophores and conidia

acinar: a small sack-like dilatation, particularly used in reference to glands

adenoma: a benign epithelial tumor

adnexal: adjunct parts

angiocarpous: a fruit body that is closed until the spores are mature

agony: pertaining to the moments just prior to death

agranulocytosis: absence of granules in cells in cytoplasm

aleukia: absence of leukocytes from the blood

anemic: a below average number of erythrocytes

anencephaly: congenital absence of the cranial vault

anophthalmia: absence of eyes

anorexia: loss of appetite

anotia: absence of ears

aplastic: anatomically underdeveloped

apothecium: a cup or saucer-like ascocarp

ascigerous: having asci

ascites: accumulation of serous fluid in the abdomen

ascocarp: a sporocarp, or ascus-producing structure in the Ascomycetes

ascogonium: the cell or group of cells in Ascomycetes fertilized by a sexual act

ascospore: a sexual spore produced in an ascus

ascus: a sac-like cell of the perfect state of an Ascomycete, in which ascospores (generally eight) are produced by free cell formation

ataxia: loss of muscle coordination

atelectasis: incomplete expansion of the lungs at birth; collapse of adult lung

atony: lack of normal tone or strength

auricular: pertaining to the ear

axillae: armpit with associated parts

axillary: pertaining to the space between the chest and extremity

basidiocarp: the basidia-producing fruit body in the Basidiomycetes

basidium: the structure on which the "sex" spores in Basidiomycetes undergo development

biliary duct: passages for conveyance of bile in and from the liver

blastic: referring to an immature stage of a developing cell

brachycardia: heart beat reduced below normal

buccal: pertaining to the cheek (oral cavity)

bullae: a large vesicle, 2 cm or more in diameter

cachetic: general ill health and malnutrition

canaliculus: an extremely narrow tubular passage or channel

canthus: angle between eyelids

catarrhal: inflammation of the mucous membranes with a free discharge

cecum: first part of the large intestine

chlamydospores: a thick-walled nondeciduous, intercalary, or terminal asexual spore made by the rounding up of a cell or cells

cholangitis: inflammation of the bile duct

cirrhosis: disease of the liver characterized by excessive fibrosis

cleistothecium: a fruit body having no special opening

clonic: rapid muscle contraction and relaxation

colic: acute abdominal pain, pertaining to the colon

conidial: pertaining to any asexual spore

conidiophore: a simple or branched hypha on which conidia are produced

conidium: any asexual spore

conjuctival sac: pertaining to membranes of the eyes

cortical: pertaining to the outer layer of an organ

crypt: blind pit or tube on a free surface

crypt epithelium: epithelial lining of a crypt

cyanosis: bluish discoloration of mucous membranes and skin resulting from a
    reduction in blood hemoglobin

desquamation: shedding of epithelial elements

dilatation: stretching beyond normal limits

Disse's space: small spaces that separate the sinusoids of the liver from liver cells

dyspnea: difficult or labored breathing

dystrophic: a disorder resulting from malnutrition

ecchymotic: a small hemorrhagic spot, larger than a petechiae

ectopia cordis: congenital displacement of the heart outside the thoracic cavity

edema: accumulation of excess fluids in intercellular spaces

emesis: vomiting

emphysema: a pathological accumulation of air in tissues, especially the lung

enteric: pertaining to the intestines

epilated: removal of hair by the roots

epistaxis: hemorrhage from the nose

erumpent: bursting through the surface of the substratum

erythema: redness of the skin caused by congestion of the capillaries

exencephaly: formation of the brain exterior to the cranium

exophthalmos: abnormal protrusion of the eyeball

extravasation: discharge or escape of blood from a vessel into the tissues

fascicle: a little group or bundle

fibrosis: fibrous tissue formation

gall: a swelling or outgrowth produced by a plant as the result of attack by a fungus
    or other organism

gastritis: inflammation of the stomach

gavage: forced feeding into the stomach

gliosis: an excess of neuroglial cells in damaged areas of the central nervous
   system

glomeruli: coils of blood vessels, especially in the kidney

glucosuria: increased glucose level in the urine

goiter: enlargement of the thyroid gland

gyri: convolution of the surface of the brain

hematopoiesis: formation of blood cells

hematuria: blood in the urine

hemorrhagic diathesis: a predisposition for hemorrhaging

hepatic: pertaining to the liver

hepatocytes: a parenchymal liver cell

hepatomegaly: enlargement of the liver

hilus: a depression or pit at that point of an organ where the vessels or nerves enter

histiocytic cells: large phagocytic interstitial cells forming part of the reticuloendothelial system

humoral: pertaining to fluid substances or those anatomical parts that are associated
   with these substances

hyaline: translucent

hydrocephalus: enlargement of the head caused by an abnormal amount of fluid in
   the cranial vault

hydropic: pertaining to an accumulation of fluid

hymenium: the sporing layer of a fruit body

hyperalfaglobulinemia: increase of α-globulins in the blood

hyperemia: engorgement of blood

hyperplasia: an abnormal increase in cell number

hyperpnea: abnormal increase in depth and rate of respiratory movements

hypha: one of the threads of a mycelium

hypoalbuminemia: abnormal low albumin content in the blood

hypoglycemia: abnormally low levels of glucose in the blood

hypophysectomy: surgical removal of the pituitary gland

hypoplastic: incomplete development of an organ so that it fails to reach adult size

hypopyon: accumulation of pus in the anterior chamber of the eye

hypothermia: low body temperature

hypoxia: deficiency in oxygen

icterus: jaundice

induration: the process of hardening

infarct: localized areas of necrotic tissue resulting from sudden deprivation of their blood supply

interstitial: interspace of tissue

intima: innermost structure

intracisternally: within a closed space serving as a reservoir for lymph or other body fluid

intrasinusoidal: within a sinus-like cavity

intussusception: the prolapse of one part of the intestine into an immediately adjoining part

ip: intraperitoneal

jaundice: pertaining to the excessive amount of bilirubin in the blood and the deposition of bile pigment in the skin and mucous membrane resulting in a yellow appearance

karyolysis: swelling cell nuclei with loss of chromatin

karyorrhexis: rupture of the cell nucleus

keratitis: inflammation of the cornea

Kupffer cells: liver phagocytes

kyphosis: hunchback

lacrimation: secretion of tears

lamina propria: a thin flat plate or layer, or the connective tissue of a mucous membrane just deep to the epithelium and basement membrane

leukocyte: white blood cell

leukocytosis: a transient increase in the number of leukocytes in the blood

leukopenia: reduction in the number of leukocytes

loop of Henle: a U-shaped turn in the medullary portion of the renal tubule

lymphocyte: mononuclear leukocyte

lymphopenia: reduction in the proportion of lymphocytes in the blood macrophage: large phagocytic cell

mast cells: connective tissue cell

medulla: the innermost part of the brain, central portion; innermost part of an organ

meningeal: pertaining to membranes enclosing the brain and spinal cord

merosporangium: a cylindrical outgrowth from the swollen end of a sporangiophore

mesentery: pertaining to membranes attaching body organs to the body wall

metulae: sporophore branches with phialides

microcephaly: abnormal smallness of the head

microvilli: minute cylindrical processes on the free surface of a cell

micturition: urination

miosis: constriction of the pupil

mucosa: mucous membrane

myelin: the lipid substance forming a sheath around certain nerve fibers

myeloid: pertaining to the bone marrow

myocardial: pertaining to the middle and thickest layer of the muscular heart wall

myxomatous: denotes relationship to mucus or slime — of the nature of a tumor comprised of primitive connective tissue cells and stroma resembling mesenchyma

nares: nostrils

necropsy: examination of a body or carcass after death

nephritis: inflammation of the kidney

nephropathy: disease of the kidney

neutrophils: a granular leukocyte

nictitating membrane: a membrane of animals which invests the nasal aspect of the globe of the eye; usually barely visible at the medial canthus

nuclear pyknosis: shrinking of nuclei and condensation of chromatin

oligodendroglial: pertaining to tissue composed of non-neural cells of the central nervous system

omentum: a fold of peritoneum extending from the stomach to adjacent organs in the abdominal cavity

opisthotonus: muscle spasms wherein the head is bent backwards

oropharynx: that portion of the pharynx lying between the soft palate and the epiglottis

ostiole: any pore by which spores are freed from an ascigerous or pycnidial fruit body

papillomas: benign tumor derived from the epithelium

paraphysis: a sterile hyphal element in a hymenium

parenteral: not through the alimentary canal, i.e., by subcutaneous, intramuscular or intravenous injection

paretic: incomplete paralysis

parietal: pertaining to the walls of a cavity

pelage: the hairy coat of mammals

percutaneous: through the skin

perithecial: pertaining to perithecium

perithecium: a flask-like ascocarp of the Pyrenomycetes usually having an ostiole

peritoneum: serous membrane lining the abdominal-pelvic walls and including the viscera

peroral: through the mouth

per os: by mouth

petechial: pinpoint spots

phagocyte: any ingesting cell

phialide: a single celled, flask-like structure from the end of which conidia are produced

phlebitis: inflammation of a vein

piloerection: erection of the hair

pinnae: protruding portion of the ear

plectenchyma: a thick tissue formed by hyphae becoming twisted and fixed together

pleural: pertaining to the serous membrane lining the lungs and thoracic cavity

polydipsia: excessive thirst

polypnea: condition wherein the rate of respiration is increased

polyuria: excessive urination

popliteal: pertaining to the posterior surface of the knee

prescapular: in front of the triangular bone in back of the shoulder

proteinuria: increased serum protein levels in the urine

protopycnidia: a primordial structure that eventually will become an asexual fruiting structure pruritic: itching

pseudothecium: a perithecium-like fruiting structure of the Pseudosphaeriales

pulmonary: pertaining to the lung

pycnidial: pertaining to the fruit body of the Sphaeropsidales, frequently globose
   or flask-like

pycnidium: the flask-like fruit body of Sphaeropsidales

pycnospores: syn. pycniospore; a spore from a haploid fruit-body or spermagonium

renal: pertaining to the kidney

reticulocyte: a young red blood cell showing basophilic reticulum staining under
   vital stain

retropharyngeal: occurring behind the pharnyx

rhizoids: a root-like structure

sarcoma: a tumor composed of connective-like tissue

satellitosis: accumulation of glial cells around neurons, usually following damage
   or injury to neurons

sc: subcutaneous

schizogenous: formed by cracking or splitting

scleral: the tough white outcoat of the eyeball

sclerotium: a firm frequently rounded, mass of hyphae, having no spores in it

sclerotized: hardened

sepsis: contamination of the blood with microorganisms or toxins

serous: pertaining to or resembling serum

somnolence: unnatural drowsiness

sorus: a fruiting structure in fungi

spina bifida: defective closure of the bony encasement of the spine

sporangia: organs producing endogenous asexual spores

sporangiole: a small sporangium without a columella

sporangiophore: a sporophore producing a sporangium

sporangium: an organ producing endogenous asexual spores

sporocarp: a fruit body

sporodochium: a mass of conidiophores tightly placed together upon a stroma of
   hyphae

sq: subcutaneous

stenosis: narrowing or stricture of a canal or duct

sterigmata: a process from a cell that supports a spore

subcutis: subcutaneous tissue

submaxillary: beneath the upper jaw

subserosal: below the serous membrane

sulci: furrow on the surface of the brain

suspensor: a specialized hyphal cell, especially one supporting a zygospore

systolic: pertaining to the contraction of the heart

tachycardia: excessive increase in heart rate

tachyphylaxis: decreasing response to repeated administration of a drug

teliospore: the resting spore from which the basidium is produced in the rust and smut fungi

tenesmus: painful and ineffectual straining at stool or urination

tetanic: continuous tonic contraction of a muscle

thrombocytopenia: decrease in the number of blood platelets

thrombosis: development of a thrombus or clot in a blood vessel formed by coagulation of the blood

tonic: pertaining to return to normal tone

tono-clonic: convulsive twitching of muscles

trabecular: supporting or anchoring strand of connective tissue

transudate: a fluid substance that has passed through a membrane or exuded from tissue

villi (intestinal): the multitudinous threadlike projections that cover the surface of the mucosa of the small intestine

xerophytic: having the ability to grow under arid conditions

Zenker's necrosis: necrosis and hyaline degeneration of striated muscle

zygospore: a resting spore (sexual) produced by the Zygomycetes

Numbers in parentheses are reference numbers and indicate that an author's work is referred to although his or her name is not cited in the text. Underlined numbers give the page on which the complete reference is listed.

Adekunle, A. A., 119(5a), 137
Adiseshan, N., 104(56), 108
Adutskevich, V. A., 66(201), 86
Ajl, S., 148(16), 157
Akhmeteli, M. A., 69, 74
Akulova, N. S., 66(170), 84
Aldasy. P., 88(7), 89
Alekseyev, G. A., 61(84), 79
Aleshin, B. V., 62, 74
Alexandrowicz, J., 110, 115
Alisova, Z. I., 22(5), 65(122), 66, 74, 81
Allen, O. N., 128(70), 140
Allison, A. C., 95, 105
Alpert, E., 10(1), 15
Alpert, M. E., 6(2, 3, 4), 15, 16, 109, 114
Altschul, A. M., 122(7, 8), 137
Amer, B. N., 135, 143
Amla, I., 9(6), 14(6), 16
Andreyev, K. P., 30(8), 65(8), 74
Anslow, W. K., 128(78), 141
Antonov, N. A., 65(9, 103, 104), 75, 80
Arai, T., 127(58), 140
Asahi, T., 122(11), 137
Ashley, R. C., 97(25), 106
Ashworth, L. J., Jr., 149(1, 2, 3, 4), 156

Babusenko, A. M., (43), 76
Bakbardina, M. K., 52(120), 81
Bamburg, J. R., 5, 16, 36(112, 183), 37(111), 69(10, 13, 14, 112), 75, 80, 85, 132(92), 142
Baral, J., 10(35), 17, 111(12), 115
Barbier, M., 133(98), 142
Barer, G. L., 36(15), 75
Barnes, J. M., 109, 115
Barnum, C. C., 128(75), 141
Bassir, O., 119(5a), 137
Bates, Fern, 109(2), 114
Beck, M. R., (10), 16
Becroft, D. M. O., 12, 16, 111, 115
Beletskij, G. N., 22, 75
Belkin, G. S., 65(9), 75
Bennett, J. E., 102(48), 107
Bennett, P. C., 100(38), 107
Bergel, F., 129(82), 141
Berlin, M. G., 65(102), 80
Betina, V., 134(100, 101, 102), 142
Beuchat, L. R., 126(39), 139
Bhamarapravati, N., 8(46), 18
Bierer, B. W., 100(39), 107
Bilai, V. I., 30, 33, 36, 41(19, 149), 52, 54, 65, (20), 75, 83
Bilbrey, R. M., 149(10), 157
Bishop, P. E., 135(115), 143
Black, H. S., 122(7, 8), 137

Blank, F., 109, 114
Blankenship, B. R., 126(35), 139
Bonner, F. L., 134(107), 142
Börner, H., 128(67, 68), 140
Boros, D. L., 94, 95(12), 105
Bösenberg, H., 127(59), 140
Bourgeois, C. H., (10), 9(12), 10(11,
    12, 42), 11(11, 12, 33, 41, 42), 12
    (33, 42), 14(11), 16, 17, 18, 111
    (15), 115
Bousquet, J. F., 133(98), 142
Boutibonnes, P., 122(17), 125(18a),
    127(28, 53, 54, 55, 57), 138, 139,
    140
Bové, F. J., 2, 16
Boy, J., 6(24), 17
Boyd, M. R., 15(14), 16, 100(41, 42,
    43), 101(42), 107
Bradley, M. O., 133(97), 142
Brewster, T. C., 99, 107
Brian, P. W., 131, 132(88), 141
Broce, D., 134(107), 142
Brodskaja, F. P., 22(180), 54(180),
    84
Brook, P. J., 147(5), 156
Brout, S. Y., 110, 115
Brown, C. M., 149(1, 2, 3, 10), 156,
    157
Büchi, G., 15(19), 16
Buckelew, A. R., Jr., 126(46), 132
    (46), 139
Buechner, H. A., 92(8), 105
Bugyi, B., 91, 105
Bullerman, L. B., 135, 143
Bundel, A. A., 29(91, 129), 36(90),
    79, 82
Burge, W. R., 126(46), 132(46), 139
Burka, L. T., 15(14), 16
Burmeister, H. R., 5(51, 52), 18, 19,
    36, 37(23, 24), 69(99), 75, 80, 126
    (40), 127(40), 132(91), 133(91), 139,
    142
Burroughs, R., 154(24), 157
Burstein, Sh. A., 62(3), 74
Butler, W. H., 11(30), 17

Caldwell, R. W., 148(22, 34), 149
    (22), 157, 158
Campbell, C. C., 92(8), 105
Cardeilhac, P. T., 97(24, 26), 106
Carll, W. T., 22(41), 76, 100(39),
    107
Castleman, B., 101(46), 107
Cernov, K. S., 69(1), 74
Chakravarti, A., 126(46), 132(46),
    139
Chandavimol, P., 9(12), 10(12), 11
    (12, 15, 41), 16, 18, 115
Chen, J.-K., 135(117), 143
Chernikov, E. A., 22(25), 75
Chernyak, B. I., 62(3), 74
Chilikin, V. I., 22(26, 27, 28), 25
    (28), 54(26, 27, 28), 55, 56(27,
    28), 59(28), 60(28), 61(26, 27,
    28), 63(27, 28), 75, 76
Chin, O., 109(3), 114
Christensen, C. M., 15(19), 16, 109,
    114, 148(6), 150(6), 156
Ciegler, A., 112(23), 115, 125(23),
    126(37, 41), 127(47, 48, 49, 60),
    136, 138, 139, 140, 143, 148(16),
    157, 163(5), 170
Clements, N. L., 126(42, 43), 139,
Cole, R. J., 126(34, 35), 133(94),
    134(108), 138, 139, 142, 147(27),
    158
Collins, D. N., 11(15), 16
Colwell, W. M., 5(52), 19, 97(24, 25,
    26), 106
Comer, D. S., (10), 10(11), 11(11),
    12(11), 14(11), 16
Conn, J. E., 128(79), 129(79), 134
    (79), 141
Converse, H. H., 150(7), 156
Cotton, R. B., 10(11), 11(11, 33),
    12(11, 33), 14(11), 16, 17
Cozas, B., 135(113), 143
Crisan, E. V., 119(5), 124, 125(5),
    137
Cuthbertson, W. F. J., 109, 115
Cutler, H. G., 133(94), 134(108), 142

Czachor, M., 110(11), 115

Damodaran, C., 135(122), 143
Damodaren, V. N., 97(22), 106
Daoud, E., 112(21), 115
Dashek, W. V., 122, 134(110), 137, 143
Davidson, C. S., 6(2, 3, 4), 15, 16, 109, 114
Davis, D. E., 122(9), 137
Davis, E. E., 134(108), 142
Davis, N. D., 148(8), 156
Davis, R. P., 125(21), 138
Davydova, V. L., 29, 52(120), 65(29, 123), 76, 81
Davydovskij, E. B., 54(30), 59(30), 61(30), 76
Dawkins, A. W., 131(88), 132(88), 141
Degurse, P. E., 36(112), 37(111), 69 (112), 80
de Serres, F. J., 126(33), 138
Detroy, R. W., 126(33a), 138, 163 (5), 170
Dickens, F., 99, 106
Diener, U. L., 148(8), 156
Dolimpio, D. A., 110(7), 115
Doupnik, B., Jr., 100(40), 107, 133 (94), 142, 150(9), 156
Drabkin, B. S., 30(31, 32), 36(144), 72(144), 76, 82
Dubuisson, H., 100(37), 107
Durston, W. E., 135(114), 143
Dvorackova, I., 12, 16
Dzhilavyan, H. A., 88, 88

Edds, G. T., 97(24), 106
Eisenberg, H. W., 101, 107
Eka, O. U., 127(52), 139
El Hassan, A. M., 111, 112(21), 115
El-Khadem, M., 122(14, 15), 134(109), 137, 142
Elliot, J. I., 154(15), 157
Ellis, J. J., 36(24, 57, 197, 203, 204), 37(24, 57, 197, 203, 204), 75, 77,

[Ellis, J. J.]
85, 86, 98(29), 99(29), 106
Ellis, J. R., 128(65), 140
Elpidina, O. K., 22(34, 35), 29(37, 38, 39, 40), 41(33), 76
Eltayeb, A. A., 112(21), 115
Enomoto, M., 5(37, 48), (38), 18, 36(198, 199), 37(198, 199), 72 (198), 85, 86
Eppley, R. M., 6, 16
Epstein, S. M., 97(23), 106
Evans, H., (10), 10(11), 11(11), 12 (11), 14(11), 16
Eyngorn, E., 62(2), 74

Fadeyeva, S. V., (200), 86
Fennell, D. I., 148(18), 149(18), 157
Feron, V. J., 99(34), 107
Fiedoruk-Poplawska, T., 102, 104 (49), 107
Fink, J. N., 92(11), 105
Fitte, J., 100(37), 107
Fok, R. A., 29(124, 125), 52(119, 120), 54(208), 65(118, 124, 125, 126), 81, 86
Ford, S., 94(13), 99(13), 105
Forgacs, J., 22(41), 76, 87, 88, 89
Franke, W. W., 133(96), 142
Frayssinet, C., 6(24), 17
Friedman, L., 94(13), 99(13), 105
Friedman, M. Yu., 22(42), 70(42), 76
Fronczak, B., 103(51), 107

Gaál, L., 88(7), 89
Gajdusek, D. C., 87, 88, 89
Gambogi, P., 136(126), 143
Gandevia, B., 104(56), 108
Ganesan, M. G., 135(122), 143
Gardner, D. E., 149(10), 157
Garren, K. H., 149(11), 157
Gäumann, E., 128(69a), 140
Geiger, B., 128(79), 129(79), 134 (79), 141

Geimberg, V. G. , (43), 76
Geminov, N. B. , 22(44, 45), 23(45),
    51, 54(45), 55(44, 45), 76, 77
Genkin, A. , 54(46), 59(46), 77
Georg, L. K. , 92(8), 105
Gerasimova, P. A. , 66(170), 84
Gerlach, W. , 37(48), 77
Germanov, A. I. , 58(47), 77
Getsova, G. , 30, 77
Gibson, J. B. , 7(43, 44), 12(44), 18
Gilgan, M. W. , 36, 69(50), 77
Gilliver, K. , 128(77), 134(77), 141
Glick, T. H. , 12, 16
Glinsukon, T. , 15, 16
Goeta, I. E. , 110(8), 115
Goldblatt, L. A. , 98(30), 106
Goldfredsen, W. O. , 36, 69(51), 77
Gopalakrishna, G. S. , 9(6), 14(6), 16
Gopalan, C. , 9(53), 19
Gordon, J. E. , 7(45), 8(46), 12(45),
    18
Gordon, W. L. , 37(52, 53, 54), 77
Gorodijskaja, R. B. , 22(55), 77
Goulden, M. L. , 148(28), 158
Gowing, N. F. C. , 101, 107
Graniti, A. , 119(1), 137
Grant, D. W. , 99, 107
Grezin, V. F. , 87(6), 89
Griffin, G. J. , 149(11), 157
Grigsby, R. D. , 147(27), 158
Grinberg, G. I. , 22(58), 54(58), 77
Grodner, R. M. , 134(107), 142
Gröger, D. , 147(12), 157
Gromashevskij, L. V. , 61(56), 77
Grossman, R. A. , 9(12), 10(11, 12),
    11(11, 12, 33), 12(11, 33), 14(11),
    16, 17
Grossmann, F. , 122(14), 137
Grove, J. F. , 36(51), 69(51), 77,
    131(88), 132(88), 141
Grove, M. D. , 36(197), 37(57, 197),
    77, 85
Gubarev, E. M. , 36(59), 77
Gubareva, N. A. , 36(59), 77
Guenzi, W. D. , 127(61), 140
Gurewitch, Z. A. , 60(60), 77
Guseva, N. A. , 69(1), 74

Gyllenberg, H. G. , 126(45), 139

Hall, H. H. , 125(23), 127(47, 49),
    138, 139
Halver, J. E. , 8(20), 16
Halweg, H. , 103(51), 107
Hamdi, Y. A. , 134(109), 142
Hamicki, J. , 135, 143
Hamilton, P. B. , 5(51, 52), 18, 19,
    97(25), 106
Hamlin, I. M. E. , 101, 107
Hansen, H. N. , 37, 71(185), 85
Harikul, S. , (10), 11(33), 12(33), 16,
    17
Harp, A. R. , 96(18), 106
Harris, C. C. , 99(32), 106
Harris, T. M. , 15(14), 16, 100(42),
    101(42), 107
Harwig, J. , 135(117), 143
Hayes, A. W. , 129(84), 131(84), 141
Heatley, N. G. , 136(124), 143
Hein, H. , Jr. , 147(27), 158
Helgeson, J. P. , 133(93), 142
Hemming, H. G. , 131(88), 132(88),
    141
Henry, M. C. , 99(32), 106
Herbst, E. J. , 126(44), 139
Herth, W. , 133(96), 142
Hesina, A. Ja. , 69(1), 74
Hesseltine, C. W. , 36(24), 37(23,
    24), 75, 98(29), 99(29), 106,
    126(33a, 40), 127(40), 132(91),
    133(91), 138, 139, 142, 148(28),
    158
Hodges, T. O. , 150(7), 156
Hoewitz, W. , 170(13), 171
Holma, B. , 96, 106
Hrootski, E. T. , 29(61), 77
Huisingh, D. , 132(89), 141
Hutt, M. S. R. , 6(2, 3, 4), 15, 16
Hyde, M. B. , 154(13), 157

Igarasi, S. , 110(8), 115
Ikawa, M. , 125(21, 22), 126(46),
    129(22), 131(22), 132(22, 46),
    138, 139

Ishii, K., 5(48), 18, 36(198, 199), 37(198, 199), 72(198), 85, 86
Ishiko, T., 110(8), 115
Ito, T., 127(58), 140
Iyengar, M. R. S., 135, 143

Jaag, O., 128(69a), 140
Jacobson, C., 110(7), 115
Jacquet, J., 122(17), 125(18a), 127 (28, 53, 54), 138, 139, 140
Jayaraj, A. P., 9(5, 6), 14(6), 16
Jayaraman, A., 126(44), 139
Jenkins, F. P., 163(4), 170
Jennings, A., 102(48), 107
Joffe, A. Z., 4(21), 17, 21(64, 68, 69, 70, 71, 72, 73), 23(71, 72, 73), 25(71, 72, 75), 26(71, 72, 73), 27 (68, 69, 73), 28(68, 69, 72, 73), 29(61, 68, 69, 73, 75), 30(31, 32, 68, 69, 71, 72, 74, 75, 77, 79, 172), 31(68, 69), 33(68, 69), 36 (73, 144, 145), 37(68, 70, 71, 72, 73, 75, 76), 41(64, 68, 69, 70, 71, 72, 74, 75, 80, 127), 42(66, 67, 68, 69, 71, 79), 44(71), 45(70, 71, 73, 75), 46(68, 71, 73, 75), 51 (70), 52(63, 68, 69, 71, 72, 74, 120), 53(64, 68, 69, 70, 73, 75, 77), 54(77), 65(9, 68, 69, 70, 71, 72, 73, 74, 75, 78, 79, 127, 128), 66, 69, 72(75, 76, 80, 144, 145, 212, 213), 74, 75, 76, 77, 78, 79, 81, 82, 84, 86, 110(9), 115, 122, 136(129), 137, 143, 148(14), 157
Johnsen, D. O., 9(12), 10(12, 42), 11(12, 42), 12(42), 16, 18
Jones, G. M., 154(15), 157
Jones, H. C., 122(8), 137
Jones, H. E. H., 99(33), 106
Jones, O. H., 100(40), 107
Just, G., 109(3), 114

Kadis, S., 148, 157
Kalberer-Rüsch, M. E., 69(191), 85, 132(88a), 141

Kamala, C. S., 9(6), 14(6), 16
Kaplan, W., 92(8), 105
Karatygin, V. M., 23(81), 79
Karlik, L. N., 22(82), 54(82), 79
Kartashova, V. L., 36(146), 72(146), 83
Kasirskij, I. A., 61(84), 79
Kathirvel-Pandian, S., 135(122), 143
Kaufman, D. G., 99(32), 106
Kaufman, L., 92(8), 105
Kaufmann, H. H., 148(6), 150(6), 156
Kavanagh, F., 135(119), 136(119), 143
Keen, P., 6(22), 17
Keller, K., 136(127), 143
Kendrick, G., 100(37), 107
Kenina, S. M., 36(146), 72(146), 83
Kennedy, B. P. C., 135(117), 143
Keschamras, N., (10), 10(11), 11 (11, 41), 12(11), 14(11), 16, 18, 111(15), 115
Kestner, A. G., 54(30), 59(30), 61 (30), 76
Ketels, K. V., 99(32), 106
Kilburn, K. H., 97(28), 98(28), 106
Killebrew, R. L., 134(107), 142
Kingrey, B. W., 100(38), 107
Kinosita, R., 110, 115
Kirksey, J. W., 126(34, 35), 133 (94), 134(108), 138, 139, 142
Kitaura, Y., 15(19), 16
Klemmer, H. W., 128(70), 140
Kobayashi, Y., 104(54), 108
Koehler, B., 148(17), 157
Kokorniak, M., 135(113), 143
Kolosova, N. I., 36(83), 79
Kong, Y. M., 92(7), 105
Korneyev, N. E., 66(170), 84
Koroleva, V. P., 66(170), 84, 87 (6), 89
Korpinen, E.-L., 87(4), 89
Kost, E. A., 54(85), 79
Kosuri, N. R., 36(57, 183), 37(57), 77, 85
Kovalev, E. N., 60(86), 79
Kováts, F., Sr., 91, 105

Koyama, Y., 127(58), <u>140</u>
Koza, M. A., 61(87), <u>79</u>
Kozin, N. I., 29, <u>79</u>
Kravchenko, L. V., 96(21), <u>106</u>
Kraybill, H. F., 6, <u>17</u>, 112(24), <u>115</u>
Kretovich, V. L., 29(91, 129), 36(89, 90, 92, 93), <u>79</u>, <u>82</u>
Kudinova, G., 30(161), <u>83</u>
Kudryakov, V. T., 59(94), <u>79</u>
Kumari, S., 9(5), <u>16</u>
Kvashnina, E. S., 30, 41(95), 54, 65 (171), 66(170), <u>79</u>, <u>84</u>, 87(6), <u>89</u>
Kwolek, W. F., 148(18), 149(18), <u>157</u>
Kykel, Yu., 30(161), <u>83</u>

Lafountain, J., 133(95), <u>142</u>
Lancaster, M. C., 163(4), <u>170</u>
Lando, Ya. Kh., 22(96, 97), 54(96, 97), <u>79</u>
Langley, B. C., 149(4), <u>156</u>
Laursen, A. C., 109, <u>115</u>
Lazáry, S., 69(191), <u>85</u>, 132(88a), <u>141</u>
LeBreton, E., 6, <u>17</u>
Lechowich, R. V., 126(39), <u>139</u>
Lee, D. J., 135(115), <u>143</u>
Lee, F. D., 135(114), <u>143</u>
Legator, M., 110(7), <u>115</u>
Legator, M. S., 96, <u>106</u>, 127(51), 129 (81), <u>139</u>, <u>141</u>
Leontiev, I. A., 61(87), <u>79</u>
Leupi, H., 128(72), <u>141</u>
Levin, I. I., 70(98), <u>79</u>
Levine, B. H., 92(7), <u>105</u>
Levitt, L. P., 12(18), <u>16</u>
Likosky, W. H., 12(18), <u>16</u>
Lillehoj, E. B., 69(99), <u>80</u>, 112(23), <u>115</u>, 125, 126(37, 41), 127(47, 48, 49, 50, 60), 136, <u>138</u>, <u>139</u>, <u>140</u>, <u>143</u>, 148(18), 149, <u>157</u>, 163(5), <u>170</u>
Lilly, L. J., 122, <u>138</u>
Lin, C.-K., 9(25), 10(25), <u>17</u>
Lin, S.-S., 9(25), 10(25), <u>17</u>
Lin, T.-M., 9(25), 10(25), <u>17</u>
Lindenfelser, L. A., 69, <u>80</u>
Ling, K.-H., 9(25), 10(25), <u>17</u>
Linnik, A. B., 69(1), <u>74</u>
Linsell, C. A., 8, 13, <u>17</u>

Linville, G. P., 15(26), <u>17</u>
Linzel, U., 127(59), <u>140</u>
Llewellyn, G. C., 122, 134(110), <u>137</u>, <u>143</u>
Löser, R., 136(127), <u>143</u>
Louria, D. B., 92(9), <u>105</u>
Lovla, D. S., 22(100), 54(100), <u>80</u>
Lowe, D., 131(88), 132(88), <u>141</u>
Lozanov, N. N., 22(101), 25(101), 54 (101), <u>80</u>
Lukin, A. Ya., 65(9, 102, 103, 104), <u>75</u>, <u>80</u>
Lundborg, M., 96, <u>106</u>
Lyass, L. S., 30, 33, 41(162), 65 (162), 66, <u>83</u>
Lyass, M. A., 22(105), 54(105), <u>80</u>
Lynn, W. S., 97(28), 98(28), <u>106</u>

Ma, D. S., 125(21), <u>138</u>
McCalla, T. M., 127(61, 62), 128 (63, 64, 65), 135(116), <u>140</u>, <u>143</u>
Mackaness, G. B., 95, <u>105</u>
McKenzie, W. N., 97(28), 98(28), <u>106</u>
McMeans, J. L., 149(1, 2, 3, 10), <u>156</u>, <u>157</u>
McNeely, B. U., 101(46), <u>107</u>
Madhavan, K., 109, <u>115</u>
Mahgoub, E. S., 102(50), <u>107</u>, 111, 112(21, 22), <u>115</u>
Maisuradge, G. I., 55, 65(106), <u>80</u>
Majima, R., 122(10, 11), <u>137</u>
Malik, O. S., 112(21), <u>115</u>
Malkin, Z. I., 22(107), 54(107), <u>80</u>
Manburg, E. M., 22(108, 109), 25 (108, 109), 54(108), 55, 56(109), 59(108, 109), 60(108), 61(108, 109), <u>80</u>
Mann, R., 134, <u>143</u>
Manoilova, O. S., 29, <u>80</u>
Marasas, W. F. O., 36(11, 183), 37 (111), 69(112), <u>75</u>, <u>80</u>, <u>85</u>, 132, <u>142</u>
Marchant, R., 134(104), <u>142</u>
Marsh, P. B., 149, <u>157</u>
Martin, P., 6(22), <u>17</u>
Mascarenhas, J. P., 133(95), <u>142</u>
Mathre, D. E., 147(21), 150(21), <u>157</u>

Matus, T. , 71(113), 80
Mayer, A. M. , 125(19, 20), 138
Mayer, C. F. , 4, 5, 14(27), 17, 22
   (114, 115), 54(114, 115), 80
Mayer, V. M. , 129(81), 141
Meeker, G. M. , 125(21), 138
Melchers, L. E. , 148(20), 157
Mellin, H. , 12(18), 16
Menke, G. , 122(14), 137
Meranze, D. R. , 109(3), 114
Merkow, L. , 97, 106
Meyer, H. , 131(86, 87), 141
Mičeková, D. , 134(100, 101, 102), 142
Michaelis, M. , 136(125), 143
Miescher, G. , 128(69), 140
Milburn, M. S. , 148(18), 149(18), 157
Milošev, B. , 87(5), 89
Mirocha, C. J. , 5, 17, 36, 72(116),
   81, 109(2), 114, 132(89, 90), 141
Mironov, C. , 54(208), 86
Mironov, S. G. , 22(117), 25(117), 29
   (124, 125), 36(121), 41(127), 51,
   52(119, 120), 65(78, 118, 122, 123,
   124, 125, 126, 127, 128), 66, 74,
   78, 81, 85
Mishustin, E. N. , 29(91), 79, 82
Mitchell, F. E. , 100(40), 107
Mogunov, B. I. , 22(192), 26(192), 54
   (192), 85
Monlux, W. S. , 100(37, 38), 107
Moran, E. T. , Jr. , 154(15), 157
Morgan, G. , 10(35), 17, 111(12), 115
Morgan-Jones, G. , 148(8), 156
Mori, Z. , 122(10, 11), 137
Morrison, A. L. , 129(82), 141
Mowat, D. N. , 154(15), 157
Murashkinskij, K. E. , 30, 82
Myasnikov, A. L. , 22(131), 54(131),
   55, 72(131), 82
Myasnikov, V. A. , 22(133), 36(121,
   132, 133), 52, 81, 82

Nair, K. P. C. , 97(26), 106
Nakhapetov, M. I. , 22(134), 54(134),
   82
Naumov, V. A. , 22(137), 82

Nelson, G. H. , 109(2), 114
Nelson, R. R. , 132(89), 141
Nemec, P. , 134(101), 142
Nesterov, V. S. , 22(135), 25(135),
   54(135), 55, 59(135), 60(135), 63
   (135), 82
Nettelsheim, P. , 98, 106
Newberne, P. M. , 8(50), 11(30), 17,
   18
Nezval, J. , 127(59), 140
Nichols, R. E. , 36(57, 112, 183,
   197), 37(57, 197), 69(112), 77,
   80, 85
Nitta, K. , 104(53), 108
Nondasuta, A. , 7(44, 45), 12(44, 45),
   18
Norris, G. L. F. , 131(88), 132(88),
   141
Norstadt, F. A. , 127(61, 62), 128(63,
   64), 135(116), 140, 143

Odelevskaja, N. N. , 22(107), 54(107),
   80
Oettlé, A. G. , 6, 17
Ohashi, S. , 104(54), 108
O'Kelly, J. , 163(4), 170
Okuniev, A. P. , 22(137), 82
Okuniev, N. V. , 36(136), 82
Olifson, L. E. , 4, 17, 29, 36(138,
   140, 142, 143, 144, 145, 146), 52,
   72(139, 140, 141, 142, 143, 144,
   145, 146), 82, 83
Oliver, P. T. P. , 134, 142
Olson, L. C. , (10), 10(11), 11(11,
   33), 12(11, 33), 14(11), 16, 17
Ong, T. , 126, 138
Oxford, A. E. , 136(123), 143
Ožegović, L. , 87, 89

Paglialunga, S. , 8(50), 18
Palti, J. , 28, 30(79), 33, 41(80), 42
   (79), 65(79), 72(80), 78, 79, 136
   (129), 143
Papadimitriou, J. M. , 96(17), 105
Pappagianis, D. , 92(6), 105

Pardo, M. , 97(23), 106
Parker, L. L. , 149(10), 157
Parpia, H. A. B. , 9(5, 6), 14(6), 16
Pathre, S. , 5, 17, 36, 72(116), 81
Patton, A. M. , 134(104), 142
Pavlović, R. , 87(5), 89
Peckham, J. C. , 100(40), 107
Peckham, J. L. , 133(94), 142
Peers, F. G. , 8, 13, 17
Pentman, I. S. , 65, 83
Pepys, J. , 91(5), 92, 95(10), 104(5),
    105
Peregud, G. M. , 58(148), 62(148), 83
Philp, J. McL. , 163(4), 170
Philpot, F. J. , 136(124), 143
Pidoplichka, M. M. , 30, 33, 41(149),
    83
Pokrovsky, A. A. , 96(21), 106
Poljakoff-Mayber, A. , 125(19, 20),
    138
Poplawska, T. , 103, 107
Port, C. D. , 99, 106
Poznanski, A. S. , 60(150), 83
Puranik, S. B. , 147(21), 150(21), 157

Rachalskij, E. A. , 22(109), 25(109),
    55, 56(109), 59(109), 61(109), 80
Ragland, W. L. , 37(111), 80
Raillo, A. I. , 23(151), 37(151), 83
Raistrick, H. , 128(78), 141
Ramazzini, B. , 91, 105
Rambo, G. W. , 148(22), 149, 157
Randhawa, H. S. , 97(22), 106
Rao, K. S. , 109(4), 115
Rátz, F. , 88(7), 89
Reddy, V. , 9(53), 19
Rehm, H.-J. , 131(86), 134, 141, 143
Reinking, O. A. , 37, 86
Reisler, A. V. , 25(153), 54(152), 83
Reiss, J. , 119, 121(6), 122(13, 18),
    125(24, 25, 26), 126(29), 127(56),
    128(71, 73, 76), 129(3, 29, 71, 73,
    80, 83), 131(29, 71, 83, 85), 132
    (29, 73, 80), 134(111), 135(111,
    120), 137, 138, 140, 141, 143
Reye, R. D. K. , 10, 17, 111, 115

Reynolds, D. W. , 12(18), 16
Ribelin, W. E. , 36(62), 37(62), 78
Rice, R. E. , 149(3), 156
Richerson, H. B. , 95, 105
Riggs, N. V. , 36(10, 11), 69(10), 75
Riker, A. J. , 128(70), 140
Rinderknecht, M. , 129(82), 141
Robinson, P. , 9, 17, 125(20), 138
Romanova, E. D. , 22(154), 25(154),
    55, 56(154), 60(154), 83
Röschenthaler, R. , 136(127), 143
Ross, V. C. , 127(51), 139
Rozhnova, Z. I. , 23(81), 79
Rubinstein, Yu. I. , 23(159), 30, 33,
    41(155, 156, 162), 65(155, 160,
    162), 66, 83
Russell, T. E. , 149(30), 158
Ryazanov, 22(163, 164), 83

Sabad, L. M. , 69(1), 74
Saint, S. , 125(18a), 127(54), 138,
    140
St. Clair Symmers, W. , 101(45), 107
Saito, M. , 5(37), (38), 18, 69(165),
    84
Sakai, K. , 5(48), 18, 36(198, 199),
    37(198, 199), 72(198), 85, 86
Samsonov, P. F. , 91, 105
Sandhu, D. K. , 97, 106
Sandhu, R. S. , 97, 106
Sargeant, K. , 163(4), 170
Sarkisov, A. H. , 87, 89
Sarkisov, A. Kh. , 22(167, 168, 169),
    30, 33, 41(67, 68, 69), 54, 65
    (166, 167, 168, 169, 171), 66, 84
Sato, N. , 5(48), 18, 36(198, 199), 37
    (198, 199), 72(198), 85, 86
Sauer, D. B. , 150(7), 154(24), 155,
    156, 157
Savel'yev, V. F. , 148(25), 157
Sawhney, V. K. , 133(99), 142
Schiffer, Z. , 110(11), 115
Schoenhard, G. L. , 135, 143
Schoental, R. , 29, 30(172), 69, 84,
    119, 137
Schor, W. F. , 111, 115

Schroeder, H. W., 147(27), 149(4), 150, 156, 157, 158

Schuller, P. L., 170, 171

Schulz, C., 136(127), 143

Scott, P. M., 36, 84, 135(117), 143

Seabury, J. H., 92(8), 105

Seemüller, E., 25(174), 37, 84

Seeni, S., 135(122), 143

Selvam, R., 135(122), 143

Sen, B. C., 9(39), 18

Serafimov, B. M., 60(178), 84

Serafimov, B. N., 60(179), 84

Serck-Hanssen, A., 9(40), 10(1, 40), 15, 18

Sergiev, P. G., 25(175, 176, 177), 84

Seto, T., 110(8), 115

Shank, R. C., 7(43, 44, 45), 8(46), 9(12), 10(12, 42), 11(12, 41, 42), 12(42, 44, 45), 15(19), 16, 18, 111, 115

Shanmugasundaram, S., 135(122), 143

Shannon, G. M., 127(50), 139

Sharbe, E. N., 22(192), 26(192), 54 (192), 85

Shepard, T. H., 15(26), 17

Sheridan, A., 163(4), 170

Shimkin, M. B., 6, 17, 109(3), 114

Shklovskaja, R. S., 22(180), 54(180), 84

Shotwell, O. L., 98(29), 99(29), 106, 127(50), 139, 148(28), 158

Shove, G. C., 153(29), 158

Sidransky, H., 97(23), 106

Siegel, N., 133(93), 142

Signer, E., 69(191), 85, 132(88a), 141

Simmons, D. G., 97(25), 106

Simonov, E. N., 65(9), 75

Simonov, I. N., 65(103, 104), 80

Simpson, J., 104(56), 108

Sinnhuber, R. O., 135(115), 143

Sirotinina, O. N., 22(181), 30, 84

Skolko, A. J., 128(70a), 140

Skripkina, Z. G., 29, 36(93), 79

Slonevskin, S. I., 22(182), 84

Slowatizky, I., 125(20), 138

Smalley, E. B., 5(7), 16, 36(11, 12,

[Smalley, E. B.] 50, 62, 112), 37(62, 111), 69(50, 112), 75, 77, 80, 85, 132(92), 142

Smirnova, V. A., 22(184), 54(184), 62(184), 85

Smith, G., 128(78), 141

Smith, T. J., (10), 10(11), 11(11, 33), 12(11, 33), 14(11), 16, 17

Smyk, B., 110(11), 115

Snyder, W. C., 37, 71(185), 85

Soboleva, P., 54(208), 86

Soboleva, R., 52(119), 81

Somers, E., 36, 84

Sosedov, N. I., 36(92), 79

Spector, W. G., 96(17), 105

Sreenivasamurthy, V., 9(5, 6), 14 (6), 16

Srikantia, S. G., 9(53), 19

Srivastava, L. M., 133(99), 142

Stähelin, H., 69(191), 85, 132(88a), 141

Stahl, C., 133(93), 142

Starkey, R. L., 135, 143

Stawicki, S., 135(113), 143

Steinegger, E., 128(72), 141

Stephenson, L. W., 149(30), 158

Stoloff, L., 163(6), 164(7), 168, 170 (13), 170, 171

Stott, W. T., 135, 143

Strong, F. M., 5(7), 16, 36(10, 11, 12, 50, 62, 183), 37(62, 111), 69(10, 13, 14, 50), 75, 77, 78, 80, 85, 132(92), 142

Strukov, A. I., 51, 62, 65(126), 81, 85

Stubblefield, R. D., 98(29), 99(29), 106, 127(50), 139

Subhamani, B., 7(45), 12(45), 18

Sugiyama, S., 110(8), 115

Sullivan, J. D., Jr., 125(22), 129 (22), 131(22), 132(22), 138

Svojskaja, E. D., 36(190), 85

Szabó, I., 88, 89

Szabó, P., 88(7), 89

Szakal, A. K., 98, 106

Szebiotko, K., 135(113), 143

Talayev, B. T., 22(192), 26, 54(192), 85

Tallent, W. H., 36(57, 203), 37(57, 203), 77, 86

Tamm, C., 36(51), 69(51), 77

Tanticharoenyos, P., 10(42), 11(42), 12(42), 18

Tatarinov, D. I., 22(193), 85

Tatsuno, T., 5(37, 47),(38), 18, 69 (165), 84

Taubenhaus, J. J., 148, 158

Taylor, E. E., 149, 157

Teregulov, G. N., 22(194), 61(194), 85

Tewfik, M. S., 134(109), 142

Thatcher, F. S., 136(125), 143

Thomas, J. B., 100(39), 107

Thomas, V. M., Jr., 126(46), 132 (46), 139

Thompson, O. C., 122(9), 137

Tijerina-Menchaca, A., 132(89), 141

Timonin, M. I., 128(66), 140

Tishchenko, M. A., 62, 85

Tkachev, T. Ya., 29, 85

Tomina, M. V., 60, 85

Tookey, H. L., 36(203, 204), 37(197, 203, 204), 85, 86

Tres, L. L., 97(28), 98(28), 106

Truelove, B., 122, 137

Tsareva, V. Yà, 22(101), 25(101), 54 (101), 80

Tsunoda, H., 5(48), 18, 36(199), 37 (199), 86

Tuite, J., 148(22, 32, 33, 34), 149 (22), 157, 158

Tulpule, P. G., 9(53), 19, 109(4), 115

Tung, T.-C., 9(25), 10(25), 17

Tutelyan, V. A., 96(21), 106

Ueno, Y., 5(48), 18, 36, 37(198, 199), 72(198), 85, 86, 136(130), 143

Umeda, M., 96, 106

Uoti, J., 87(4), 89

Uraguchi, K., (38), 5, 18, 104(55), 108

Uritani, I., 122(10), 137

Valentine, H. D., 100(39), 107

Van As, A., 97, 106

Vanderhoef, L. N., 133(93), 142

Vanderwoude, W. J., 133(96), 142

Vedder, J. S., 111, 115

Veindrach, G. M., (200), 86

Veress, B., 112(21), 115

Verney, E., 97(23), 106

Verona, O., 136(126), 143

Vertinskij, K. I., 66(201), 86

Vickers, C. L., 100(39), 107

Viitasalo, L., 126(45), 139

Voronin, V. M., 69(1), 74

Wallen, V. R., 128(70a), 140

Wang, F. H., 128(74), 141

Wang, J.-J., 9(25), 10(25), 17

Ward, J. L., 129(82), 141

Warren, K. S., 94, 95(12), 105

Waynforth, H. B., 99(33), 106

Webster, D. R., 12, 16, 111, 115

Weeks, B. A., 5(51), 18

White, A. F., 119, 137

Wieder, R., 109(3), 114

Wiewiorowska, M., 135(113), 143

Wightman, R., 15(19), 16

Wildman, J. D., 125, 138

Wilson, B. G., 100(41, 42, 43), 101 (42), 107

Wilson, B. J., 15(14), 16, 111(20), 115

Wogan, G. N., 6(4), 7(43, 44, 45), 8(46, 50), 12(44, 45), 15(19), 16, 18

Wolf, J. C., 132(90), 141

Wolff, I. A., 36(57, 203), 37(57, 203), 77, 86

Wollenweber, H. W., 37, 86

Wooding, W. L., 9(12), 10(12, 42), 11(12, 42), 12(42), 16, 18

Wragg, J. B., 127(51), 139

Wright, D. E., 99(36), 107, 134(106), 142

Wright, J. M., 134(105), 142

Wu, R., 9(25), 10(25), 17

Wyatt, E. P., 129(84), 131(84), 141

Wyatt, R. D. , 5, 18, 19

Yadgiri, B. , 9, 19
Yagen, B. , 72(212, 213), 86
Yamaguchi, M. , 104(54), 108
Yamakawa, Y. , 136(130), 143
Yamamoto, I. , 104(52, 53), 108
Yamamoto, Y. , 104(53), 108
Yamasaki, E. , 135(114), 143
Yang, D. T. C. , 100(41, 42), 101(42),
  107
Yasnitskij, P. Ya. , 61(87), 79
Yates, S. G. , 36(57, 197), 37(57, 197,
  203, 204), 77, 85, 86, 148(36), 158
Yefremov, V. V. , 21(206, 207), 22

[Yefremov, V. V.]
  (206, 207), 25(207), 26, 55, 56
  (206, 207), 60(206, 207), 61(205,
  206, 207), 86
Yershova, O. A. , 29, 79
Young, R. C. , 102, 107
Yuan, S. S. , 15(19), 16
Yudenich, V. 52(119), 54(208),
  81, 86
Yuskiv, R. V. , 87(8), 89

Zavyalova, A. P. , 36(209), 86
Zhodzishkij, B. Ya. , 54(210), 86
Zhukhin, V. A. , 22(211), 86
Zuffante, S. M. , 96(18), 106

Absidia repens, 126
Actinomyces, 92
  globisporus, 33
  griseus, 31, 33
Actinomycetaceae, 92
Acute cardiac beriberi (shoshin-
    kakke), 5
Adenomatosis, 99
Aeration of grain storage bins, 152
Aerosols, 97
  of aflatoxin B_1 in animal studies, 97
Aflatoxicoses, human, 6
Aflatoxicosis, 11, 12
  in Taiwan, 10
Aflatoxin, 96, 97, 103, 109-113, 147-
    151, 155, 162-170
  action guide for, 163
  assay methods, 163
  carcinogenic effects, 99
  consumption, effect on hepatocellu-
    lar carcinoma incidence in
    Thailand, 7
  control program, 163
  in corn, 148, 149
  in cotton, 149
  effects on cellular organization, 122
  effects on enzymes, 122
  effects on nucleic acids, 122
  effects on proteins, 122
  effect on respiratory tract, 99
  in man, 1
    cause of liver cancer, 6-9
  in monkeys, 10, 11
  residues in edible tissue, milk, and
    eggs, 166
  sampling plans, 163
  -susceptible commodities, 164

[Aflatoxin]
  Brazil nuts, 164
  copra, 164
  corn, 164
  cottonseed, 164
  domestic tree nuts, 164
  pistachio nuts, 164
  tolerance, 170
Aflatoxins, 134
  collaboratively studied analysis
    methods for, 170
  effects on algae, 125
  effects on bacteria, 126
  effects on fungi, 125
  effects on higher plants, 119
  effects on seed germination, 119
  in leftover cooked foods, 8
  in peanuts, 6
  in U.S. peanut products, 163
Aflatoxin B_1, 7-12, 96, 102-104,
    110, 119, 121-127, 134
  carcinogenicity of, 6
  chronic toxicity of, 6
  effect on metabolic activity of root
    tissues, 122
  in man, acute toxicity of, 9
  in rats, 99
  studies in guinea pigs, 97, 98
  studies in hamsters, 97
Aflatoxin B_2, 11, 110, 125
Aflatoxin B_3, 126
Aflatoxin G_1, 12, 99, 110, 125-127
  in rats, 99
Aflatoxin G_2, 110, 125
Aflatoxin R_0, 134
Alimentary toxic aleukia (ATA), 1,
    4, 5, 14, 15, 21-74, 110,

[Alimentary toxic aleukia (ATA)]
    113, 148
  bioassay methods and procedures, 26
    with toxic _Fusarium_, 29
  clinical characteristics of, 54-59
    four stages, 54-59
  clinical course of, 4
  effect of meteorological conditions
    on, 49, 50
  environmental and agricultural
    factors, 26
  epidemiological background, 21
  etiology of, 41
  factors affecting outbreaks, 25, 26
  fungi, 30
  fungi and toxins, 30
  mycological study, general picture
    of, 41
  Orenburg district, 24
  other tests, 29
  outbreaks, toxicity of samples of
    cereals, plant organs, and
    soil in, 46
  pathology of organs in man, 59
  prophylaxis and treatment, 70
  reaction of skin test, 28
  review of outbreaks, 22
  in Russia, 21
  seasonal effects, 46
  skin test on rabbits, 26
  source of causal toxins, 46
  stages of development, 55-59
  symptoms of, 22
Allergy, 92
Allium _cepa_, 122, 123, 128, 129, 132
Allium _sativum_, 125
Almonds, 166
Alternaria, 30, 31, 33, 109, 110, 113,
    150
  humicola, 32
  tenuis, 32
Alveolitis, 95
  in humans, etiologic agents in, 93
American Association of Cereal
    Chemists (AACC), 163
American Oil Chemists' Society
    (AOCS), 163

Ammoniation of cottonseed meal,
    167
Analytical methods for mycotoxin
    contamination, 168
Analytical standards for mycotoxin
    contamination, 168
Antibiotics, 119
Ascochyta _pisi_, 128
Aspergillic acid, 134
Aspergillosis, 99, 101, 102
  in man, 101
Aspergillus, 91, 92, 94, 97, 101,
    109-111, 113, 127, 134
  awamori, 125
  caliptratus, 30
  candidus, 110
  flavus, 12, 96, 99, 102, 110-113,
    122, 125, 136, 149-151,
    165
    in cotton, 149
    in areas of high temperature,
      149
    in areas of low relative
      humidity, 149
    development in the field, 148
    in peanuts, 149
  fumigatus, 33, 102, 103
  glaucus, 102, 110
  nidulans, 134
  niger, 33, 104, 125, 128, 129,
    148, 150
  oryzae, 110, 128
  parasiticus, 125
  versicolor, 136
Association of Official Analytical
    Chemists (AOAC), 163, 164
  associate referees, 164
Asthma, oosponol implicated in, 104
Avena _sativa_, 133, 134
Avitaminosis, 5

Bacillus
  brevis, 126
  cereus, 127, 134
  enteritides, 136
  licheniformis, 127

[Bacillus]
  megaterium, 126, 127, 131, 132
    NRRL 1368, 135
    sensitivity to aflatoxin B₁, 135
  mycoides, 127
  stearothermophilus, 127
  subtilis, 128, 131, 132, 134, 135
  thuringiensis, 127
Bermuda grass tremors, 148
Blastomyces, 92
Blastomycosis, 91
Botrytis cinerea, 33, 134
Brassica oleracea var. botrytis, 128
Brazil nuts, aflatoxin contamination in,
    165
Brevibacterium, 127
Butenolide, 37, 72, 133
  toxic, 133
Byssochlamic acid, 131

Cancer, liver, in man, 1
Candida, 92, 134
  albicans, 109, 136
  parapsilosis, 109
Caulerpa prolifera, 133
Cell cultures for in vitro studies of
    mycotoxins, 96
Chaetomium, 109
  brefeldi, 33
  trilaterale, 134
Chlorella pyrenoidosa, 125, 129, 132
Citrinin, 134-136
Cladosporium, 31, 33, 51, 110, 113,
    150
  epiphyllum, 32
  exoasci, 32
  fagi, 32
  fuligineum, 32
  gracile, 32
  graminum, 33
  gramosum, 33
  herbarum, 32, 125, 129
  molle, 32
  penicillioides, 32
  pisi, 32
  spherospermum, 33

Claviceps
  paspali, 2
  purpurea, 2, 147
Clavine alkaloids, 2
Coccidioides, 92
Coccidioidomycosis, 91
Cochliobolus
  carbonum, 132
  heterostrophus, 132
Coremium glaucum, 33
Corn, 149
  aflatoxin contamination in, 149,
    165, 166
  contamination by A. flavus, 150
Corynebacterium rubrum, 134
Cotton, 149
  aflatoxin in, 149
Cottonseed, 150, 166
  aflatoxin contamination in, 166
  contamination by A. flavus, 150
Cryptococcus, 92
Cucumis sativa (cucumber), 122
Cyclochlorotine, 96
Cytochalasin (s), 133, 134
  A, 134
  B, 15, 133, 136
  D, 134
  E, 15
  effect on higher plants, 133

Dactylium dendroides, 126
Detoxification of contaminated
    products for mycotoxin con-
    trol, 167
Deuteromycetes, 92
Diacetoxyscirpenol, 36, 69, 131

Endothia parasitica, 136
Ensiling, 154
Epicladosporic acid, 110
Epidemiological design in human
    mycotoxin studies, 13
Epidemophyton flocossum, 109
12,13-epoxytrichothecene, 5
Ergoline, 2

Ergot alkaloids, 2, 15
  derived from lysergic acid, 3
    ergocornine, 3
    ergocristine, 3
    α-ergokryptine, 3
    β-ergokryptine, 3
    ergosine, 3
    ergostine, 3
    ergotamine, 3
  structure of, 2
Ergotism, 1, 2, 147
  in man, 1
Escherichia coli, 127, 131, 132, 134,
    136
Euglena gracilis, 125

F-2 (zearalenone), 132
Facial eczema, 147
Fagiocladosporic acid, 110
Farmer's lung, 92, 104
Federal Food and Drug Administration
      (FDA), 161, 163-170
  regulatory programs, 164
Federal register, 170
Fescue foot, 148
Flavobacterium aurantiacum, 127
Food and Agriculture Organization
      (FAO), 170
Food, Drug, and Cosmetic Act, 161
Fungal cultures and cereal samples,
      comparison between the
      toxicity of, 51
Fungal toxicity, growth characteristics
      associated with, 54
Fungi in overwintered cereal crops,
      30-33
Furanoterpenoids, 15
Fusariogenin, 110
Fusarium, 4, 26, 31, 33, 36, 51, 52,
      54, 65, 69, 72, 74, 109, 110,
      113, 150, 155
  arthrosporioides, 34
  avenaceum, 34
  conditions for toxin formation, 42
  culmorum, 34, 128
  equiseti, 34, 131

[Fusarium]
  var. acuminatum, 34
  var. caudatum, 34
  fungi, 34, 52
    biological properties of, 52
    toxicity of from overwintered
        cereals, 34
  graminearum, 34
  invasion of corn in wet harvest,
      148
  isolation of, 42
    materials and methods, 42
  javanicum, 34, 99
  lateritium, 30, 34
  moniliforme, 34, 133
  mycotoxins, 132, 136
  nivale, 34
  oxysporum, 34
  poae, 5, 21, 27-30, 33, 34, 36-42,
      44, 45, 51-54, 59, 61-73
  roseum "graminearum", 132
  sambucinum, 34
  semitectum, 34
  solani, 34, 99
  sporotrichiella section, 21, 27,
      29, 52, 54, 71
    taxonomic problems, 37-41
  sporotrichioides, 5, 21, 27-30,
      33, 34, 36-45, 48, 51-54,
      59, 61, 63-66, 69-73
    var. chlamydosporum, 37-41
    var. tricinctum, 27-30, 33, 34,
        36-41, 65, 73
  temperature in relation to stage of
      development, 53
  toxins, tests for, 29
  tricinctum, 37, 69-71, 132, 133

General referee for mycotoxins, 164
Gibberellic acid, 122
Gliocladium, 30, 33
  ammoniophilum, 30
  penicillioides, 33
Gliotoxin, 97
Glycine max (soybean), 132
Gonobotrys flava, 33

Gossypium hirsutum (cotton), 122
Grain drying, 152
Guidelines by FDA, 169

High moisture storage of grain, 153,
        154
  ensiling, 154
  using chemical preservatives, 154
Histoplasma, 92
  capsulatum, 94
Histoplasmosis, 91
HT-toxin, 37, 69
HT-2 toxin, 72
Human liver cancer, studies in
    Kenya, 8
    Swaziland, 6
    Uganda, 6
Human stachybotryotoxicosis, 87
3-hydroxymethyl keton-8-hydroxyiso-
        coumarin, 104
Hymenopsis, 30
Hypersensitivity, 92

Idiopathic disease, mycotoxins involved
        in, 169
Imperfect fungi, 92
Importer control programs, 165
Indian childhood cirrhosis, 1
Interlaboratory collaborative studies,
        163
International Union for Pure and
        Applied Chemistry (IUPAC),
        164
Ipomaea batatas (sweet potato), 12

Kalanchoe daigremontiana, 134, 135
Kwashiorkor, 9, 14

Lactuca sativa (lettuce), 119, 133
Lepidium sativum, 119, 120, 124, 125,
        128, 129, 133
[14C] leucine, 122
Lillium longiflorum, 122, 133, 134

Liver cancer, human, 1
Low temperature drying of grain, 153
Lung edema, 100
Luteoskyrin, 96, 136
Lysergic acid, 2
  derivatives, 2
Lysosomes, 96

Macrophages, 95
Macrosporium commune, 33
Malva rotundifolia, 128
Metaplasia, 98
  in animal studies of aflatoxin, 98
Micrococcus, 131
    pirogenes var. aureus, 132
Microsporum, 109
Moisture migration prevention, 152
Moniliformin, 133
Mortierella candelabrum
    var. minor, 30
Mortierella polycephala, 30
Mucor, 31, 33, 110, 113
    albo-ater, 30, 31, 32
    corticola, 32
    dispersus, 33
    fumosus, 33
    globosus, 33
    griseo-cyanus, 126
    griseo-ochraceus, 33
    heterosporum, 33
    hiemalis, 30, 32, 125
    humicola, 32
    mucorum, 33
    oblongisporus, 33
    racemosus, 32
    ramannianus, 133
    silvaticus, 33
Mucorales, 92
Mycelia, in aspergillosis, 101
Mycobacterium, 95
    smegmatis, 135
Mycoses, 91, 92
Mycotoxicoses of man, 1-15
    dietary considerations, 1
    epidemiological considerations, 1

Mycotoxin(s)
  contamination by, 147–156
  contamination of foods, 161, 162
  definition of, 119
  in foods and feed stuffs, 147
  in human pulmonary disease, 91–105
  laboratory assays, 14
  in man, 1
    in the hemic system, 109
  occurrence and consumption, 13, 14
  produced in the field, 147
  as protein synthesis inhibitor, 122
  residues in meat, milk and eggs, 162
  storage contamination, 150

Nasturtium officinale, 125
Necrosis in aspergillosis, 101
Neosolaniol, 37, 72
Neurospora crassa, 126
Neurotoxic mycotoxins, 5
β-nitropropionic acid, 111
Nitella, 133
Nivalenol, 96
Nocardia, 92, 127

Ochratoxin(s), 134, 168
  A, 97, 136
Official methods of analysis, 164
Oosponol, fungal metabolite, 104
Oospora astringenes, 104
Oosporein, 134
Organ cultures for in vitro studies of
    mycotoxins, 97
Overwintered cereal crops, fungi and
    toxins in, 30-33, 41-43, 45-
    48, 51, 52
Overwintered grain(s), 5, 110, 148
  F. poae and F. sporotrichioides in,
    59-69
    effect in animals, 63-69
    effect on various human organs,
      59-63
  Fusarium toxin in, 70, 71, 74
  toxins, 36
Overwintered millet grain, toxicity at
    different seasons, 48

Paecilomyces varioti, 131
Patulin, 96, 97, 127-129, 135, 168
  effect on bacteria, 129
  effect on fungi, 128
  effect on higher plants, 127
Peanut Administrative Committee,
    165
Peanuts, 149, 165
  aflatoxin in, 149
  aflatoxin contamination in, 165
Pecan, 166
Penicillic acid, 96, 134, 136, 168
Penicillium, 31, 33, 109-111, 113,
    127, 134, 148, 150, 155
  auratio-virens, 33
  biourgeianum, 33
  brevi-compactum, 32, 136
  chrysogenum, 32, 125
  citreo-roseum, 33
  citreo-viride, 136
  citrinum, 136
  crustosum, 33
  cyaneo-fulvum, 33
  cyclopium, 32, 136
  digitatum, 133
  duclauxi, 125
  expansum, 125, 128
  griseo-roseum, 33
  houardii, 33
  martensi, 33
  micynskii, 33
  nigricans, 32
  notatum, 32
  palitans, 33
  purpurogenum, 33
  steckii, 32
  umbonatum, 32
  urticae, 127, 136
  viridicatum, 32
  westlingi, 33
Peptide-type alkaloids derived from
    lysergic acid, 3
  ergine, 3
  ergometrine (ergobasine), 3
  lysergic acid-L-valinemethylester,
    3
  lysergic acid methylcarbinol-
    amide, 3

Pestalotia, 110
Phagocytosis, 95
Phoma, 30
Photobacterium fischeri, 135
Phytophthora erythroseptica, 128
Phytotoxins, 119
Piptocephalis freseniana, 30-32
Pistachio nuts, aflatoxin contamination
    in, 165
Pisum sativum, 121, 128
Pithomyces chartarum, 147
Poaefusarin, 5
Polyporus biennis, 134, 136
Potato-dextrose-agar (PDA), 38, 40,
    42
Preservative for grain storage, 155
Prevention of mycotoxins in food
    products, 147
Primary liver cancer, 6
Propionic acid, 154, 155
Prosomillet, 4, 30, 41
    toxin in glumes, 52
Proteus vulgaris, 127
Psuedomonas aeruginosa, 127
Pulmonary clearance, 95
Pulmonary disease, 91
Pulmonary toxicity by mycotoxins, 96
    animal studies, 97-101
    in vitro studies, 96, 97
    man, 101-104
Pulmonary zones affected by fungi, 93
    alveoli, 94
    tracheobronchial tree, 93, 94
Pythium, 128

Raphanus sativus, 133
Regulatory action by FDA, 169
Reye's syndrome (disease), 1, 10-14,
    110, 111, 113
    in Czechoslovakia, 12
    in New Zealand, 12
    in Thailand, 13
Rhizopus, 92, 126, 128
    nigricans, 31, 32, 125, 129, 131
Rhodotorula rubra, 133
R. glutinis, 133

Rubratoxin B, 96, 129, 135
    effects on algae, 129
    effects on bacteria, 131
    effects on fungi, 129
    effects on higher plants, 129

Sabouraud's agar, 103
Saccharomyces
    carlsbergensis, 133
    cerevisiae, 129, 131, 135
        effect of B_1 on fermentation
            activity, 126
    ellipsoideus, 135
    pastorianus, 133, 136
Safe storage time for grain, 151
Safe tolerances for aflatoxins, 163
Safety guidelines for mycotoxins in
        foods, 162, 168
Saint Anthony's fire, 1
Salmo gairdneri (rainbow trout), 135
Salmonella paratyphi, 136
Salmonella typhimurium, 135
Sampling, 162
Sampling plans for aflatoxins, 170
Schistosoma, 95
Sclerotium, 109
    rolfsii, 136
Scopulariopsis brevicolis, 109
Screening methods for mycotoxin
        control, 167
Shoshin-kakke (acute cardiac beri-
        beri), 5
Sinapsis alba, 131
Solanum lycopersicum, 128
Sporidesmin, 97, 134
Sporofusarin, 5
Stachybotryotoxicosis, human, 87-89
    control, 88
    symptoms, 87, 88
Stachybotrys alternans (synonyms,
        S. atra and S. chartarum),
        87
Stachybotrys atra (synonyms, S.
        alternans and S. chartarum),
        87
Stachybotrys chartarum (synonyms,
        S. atra and S. alternans), 87

Stachybotrys-contaminated straw, 87,
        88
Staphylococcus, 131
    aureus, 127, 136
Sterigmatocystin, 97, 136, 168
Storage fungi, 148, 150
Streptococcus, 127
Streptomyces, 92
Surveillance for mycotoxin contamina-
        tion, 168

T-2 tetraol, 72
T-2 toxin, 5, 36, 37, 69, 72, 74, 132,
        168
    Fusarium isolates used, 73
        60/9, 73
        396, 73
        792, 73
        958, 73
        NRRL 3287, 73
        NRRL 3299, 73
        60/10, 73
        347, 73
        351, 73
        738, 73
        921, 73
        1182, 73
        1823, 73
        NRRL 3249, 73
        NRRL 5908, 73
        2061-C, 73
        YN-13, 73
    identification tests, 72
    structure of, 4
Tetracyclic sesquiterpenoids, 69
Thamnidium elegans, 31, 32, 125, 129
Thermophilic Actinomyces, 91
Tolerance of mycotoxin in food, 168
Torulopsis utilis, 136
Toxic fungi, 32, 53
    effect of substrate on growth, 53
    isolated from overwintered cereals
        and soils, 32
Toxic glucosides isolated from proso-
        millet, 4
Toxicity
    of cereals as related to ATA inci-
        dence, 45

[Toxicity]
    of fungal cultures and cereal
        samples, comparison
        between, 51
    of overwintered millet grain, 48
    persistence of in stored grain, 45
    test on rabbits, 27, 28
Toxicology, FDA studies, 168
Toxin(s),
    conditions for formation, 42, 44-
        46, 151
        moisture contents, 151
        relative humidities, 151
    in overwintered cereal crops, 30-
        33
    of overwintered grain, 36
Toxomycoses, 91, 92
    in humans, 93, 95
        etiologic agents in, 93
Tradescantia paludosa, 133
Trichoderma, 128
    lignorum, 30, 31, 32
Trichophyton, 109
Trichothecene(s), 69, 168
    toxins, 30
Trichothecium roseum, 31, 33
"Turkey X" disease, 163

United Nations Environment Program
        (UNEP), 170
United States Department of Agri-
        culture (USDA), 164-168
    surveillance programs, 164
Ustilago tritici, 128

Verticillium lateritium, 31, 33
Vicia faba, 122
Vinca minor, 133
Volvox aureus, 129

Walnuts, 166
World Health Organization (WHO), 170
Zea mays, aflatoxin effect upon, 125
Zearalenone (F-2), 72, 132, 168